COUNTRIES AND CULTURES OF THE WORLD

ITALY

SOCIAL, ENVIRONMENTAL AND AGRICULTURAL ISSUES

COUNTRIES AND CULTURES OF THE WORLD

Additional books in this series can be found on Nova's website under the Series tab.

Additional e-books in this series can be found on Nova's website under the e-book tab.

COUNTRIES AND CULTURES OF THE WORLD

ITALY

SOCIAL, ENVIRONMENTAL AND AGRICULTURAL ISSUES

GINA PITTMAN
EDITOR

nova publishers
New York

NOTICE TO THE READER

Library of Congress Cataloging-in-Publication Data

ISBN: 978-1-63484-408-6

Published by Nova Science Publishers, Inc. † New York

CONTENTS

PREFACE

Italy's peculiar cultural variability is the result of population movements and invasions from the Mediterranean Sea, which took place in both historical and prehistoric times. Due to many interesting places such as the city of Rome, Central Italy was endured by migrants, who have contributed to Italy's culture, linguistics and also genetics. Chapter One reconstructs human demic movements in central Italy using both archeology and genetics. Chapter Two addresses issues in the conceptual context of land take, which is a change in agriculture, forestry, and other lands due to urbanization and artificial land development, in Sardinian urban areas of Italy. Moreover, global warming has had a negative impact on water balance, natural systems, and agriculture activities. Chapter Three provides a seasonal analysis of crop evapotranspiration based on the Penman-Monteith equation in the Po River Delta in Italy.

Chapter 1 – The great cultural variability in Italy is the result of population movements and invasions from all Mediterranean Sea, which took place in both historical and prehistoric times. Because of many interesting places such as the city of Rome, the Central Italy was endured by arrival of small human groups, which would carry their cultural and linguistic and also genetic contributions. A high number of the prehistoric sites and the substantial cultural diversity of those people living in all Central Italy show animated history already before Roman period. However, in the last decades, there has been a gradual interest to a new and important tool for studying the human migrations: genetics. Along with cultural anthropology and archeology, this instrument could be useful to shed light on forgotten historical events, and to unravel the genetic structure of ethnolinguistic minorities and small mountain communities. A very interesting case study is the biodemographic reconstruction of the mountain communities located between Latium and Abruzzi through the genetics. This work highlighted a genetic similarity between the Central Italian samples and Near Eastern populations mainly in the male genetic pool. In conclusion, in this Chapter the authors provide to make state-of-art on previous work that reconstruction of the human demic during the centuries through both archeology and genetics.

Chapter 2 – Land take is the “Change of the amount of agriculture, forest and other semi-natural and natural land taken by urban and other artificial land development.” In 2011, the European Commission (EC) put in evidence that an important milestone for the EU should be to reach the goal of no net land take by 2050, and to take under strict control the impact on land-taking processes of the EU policies in the new Structural Funds programming period (2014-2020) (Communication of the EC to the European Parliament COM (2011) 571 of 20.9.2011).

In this paper the authors address two intertwined issues in the conceptual context of land take. First, they analyze and compare the characteristics of a fundamental phenomenon related to land take that is the change from non-artificial to urbanized areas in medium-sized towns. The authors study the phenomenon in some coastal cities of Sardinia and try to put in evidence a few lessons, both in terms of determinants of urbanization of non-artificial land and of planning policy implications. Second, they consider one of the most important land take-related issues, that is urban renewal concerning historic centers’ preservation. Indeed, restoration and reuse-based policies of historic centers is one of the most effective ways of preventing land-taking processes, especially in terms of increasing the housing stock. Under this perspective, the authors analyze the determinants of the perceived quality of the urban environment of Cagliari’s historic district through a survey addressed to the resident families and administered through a questionnaire. The survey is interpreted through a discrete choice model whose results put in evidence several important inferences related to residential satisfaction. These inferences could be very helpful for the definition of future public policies for a substantial improvement in the quality of life of historic centres, related to public services and infrastructure endowment, for example public parking, refuse collection and retail shops.

Chapter 3 – The recent global warming has negative impact on water balance, natural systems, and agriculture activities. Assessing the crop evapotranspiration, the authors estimated the fraction of water consumed by vegetation. They have calculated the crop evapotranspiration in Po River Delta with Penman-Monteith equation using daily input data and through Budyko method the water balance was calculated. In relationship with crop growth, the authors defined five stages for 2013: initial, crop development, mid-season, late season, and the cold season. The field coefficient was calculated to estimate the evapotranspiration for entire Po River Delta. In the mid-season was observed that field coefficient has a major value of 1.341 and crop evapotranspiration values than exceed 290 mm/month were identified in summer period: 291.37 mm in June, 332.93 mm in July, and 295.19 mm in August. The major rate of potential evapotranspiration was carried out for July, with a value of 248.77 mm. With 149 mm of precipitation in summer, the first results indicate that Po River Delta had a water deficit in mid-season. The authors consider that these findings represent a quantitative analysis about evapotranspiration in the most important wetland area from Italy and could be useful for hydrologists and climatologists.

In: Italy
Editor: Gina Pittman
ISBN: 978-1-63484-408-6

Chapter 1

CENTRAL ITALY: HISTORY OF ANCIENT AND CURRENT MIGRATIONS

Francesco Messina[1,*], Mario Federico Rolfo[2] and Olga Rickards[1]
[1]Dipartimento di Biologia, Università di Roma "Tor Vergata," Roma, Italy
[2]Dipartimento di Beni Culturali, Musica e Spettacolo, Facoltà di Lettere e Filosofia, Università di Roma "Tor Vergata," Roma, Italy

ABSTRACT

The great cultural variability in Italy is the result of population movements and invasions from all Mediterranean Sea, which took place in both historical and prehistoric times. Because of many interesting places such as the city of Rome, the Central Italy was endured by arrival of small human groups, which would carry their cultural and linguistic and also genetic contributions. A high number of the prehistoric sites and the substantial cultural diversity of those people living in all Central Italy show animated history already before Roman period. However, in the last decades, there has been a gradual interest to a new and important tool for studying the human migrations: genetics. Along with cultural anthropology and archeology, this instrument could be useful to shed light on forgotten historical events, and to unravel the genetic structure of

* framessins84@libero.it.

ethnolinguistic minorities and small mountain communities. A very interesting case study is the biodemographic reconstruction of the mountain communities located between Latium and Abruzzi through the genetics. This work highlighted a genetic similarity between the Central Italian samples and Near Eastern populations mainly in the male genetic pool. In conclusion, in this Chapter we provide to make state-of-art on previous work that reconstruction of the human demic during the centuries through both archeology and genetics.

INTRODUCTION

Italy's peculiar cultural variability is the result of population movements and invasions from the Mediterranean Sea, which took place in both historical and prehistoric times. Geographically, central Italy divides the Tyrrhenian Sea from the Adriatic Sea, while the Apennines Mountains (located in Abruzzi, Molise, Umbria, Tuscany, and the Southeastern Latium regions) have the highest peaks and the most rugged terrain of the Italian peninsula, excepting the Alps mountain chain (Figure 1). A high number of the prehistoric sites in central Italy show an animated history between the late Pleistocene phase and the pre-Roman period: in fact, the region has been inhabited since at least the Lower Paleolithic period, as testified by *H. cepranensis* (Muttoni et al., 2009). And from the Mesolithic to the Neolithic and then Bronze age periods various cultures have arisen in this area, as evidenced by the Villanovan culture and Etruscans (Thuiller, 2008). During the historical period, because of its central position and proximity to Rome, central Italy was traversed by travelers from throughout the Mediterranean Sea area, who carried their cultural identity. However, during the Middle Ages the presence of numerous small groups speaking Slavic-languages along the Italian peninsula suggested a migration route from the northeastern Mediterranean Sea, likely near the Balkans and the Anatolian peninsula (Babalini et al., 2005, Boattini et al., 2011). Migrations in Italy and all Europe have always been investigated through cultural and archeological disciplines, but in recent decades a new and important tool has gained acceptance: genetics. Along with cultural anthropology and archeology, this powerful instrument could shed light on forgotten historical events and unravel the genetic structure of ethnolinguistic minorities or small mountain communities. In small human groups the cultural effects, migration patterns and single historical events may leave important signatures on the genetic pools (Trupiano, 2012), and, following this indication, the many linguistic minorities have been broadly analyzed from both a biodemographic and a genetic point of view. These linguistic minorities include the Provençal of the Italian Alps; the Croatian communities in Molise; and the Arbereshe, an Albanian-speaking ethnic minority, (Babalini et al., 2005, Biondi et al., 2005, Boattini et al., 2011, Tagarelli et al., 2007, Boattini et al., 2010, Capocasa et al., 2014).

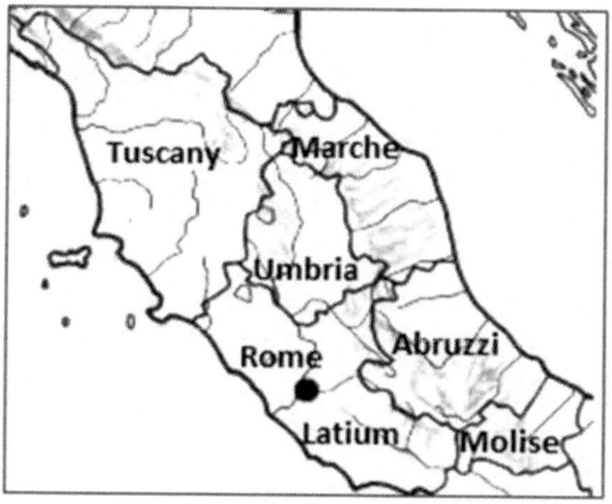

Figure 1. Geographical map showing central Italy.

Human DNA could be used to investigate human population history and evolutionary studies, from a single genetic variant to the whole human genome. However, in the last twenty years the most exploited have been the Y chromosome and mitochondrial DNA (mtDNA), because of their uniparental inheritance (only maternal for mtDNA and paternal for the Y chromosome) and their lack of recombination (Jobling & Tyler-Smith, 2003, Soares et al., 2010).

In this chapter we reconstruct human demic movements in central Italy using both archeology and genetics. We describe cases of the central Italy, in which genetics resolved doubts about origins. These cases highlighted a genetic similarity between the central Italian samples and the near eastern populations. The cases to be discussed are as follows: the state-of-art of genetics research on the Etruscan people in Tuscany; evaluation of the maternal genetic diversity of the Croatian-speaking minority in Molise and its genetic relatedness to neighboring Italian groups and the Croatian parental population; and the historical reconstruction of the mountain communities located in the Upper Aniene Valley between Latium and Abruzzi through the genetics.

ARCHEOLOGICAL BACKGROUND

During the Holocene phase the central Italian regions showed periods of demographic instability that continued until the Neolithic period. In many locations, human bones were found in archeological contexts: in Latium, in Mora Cavorso (a cave near Jenne, Rome) a human presence could be dated to the Neolithic period. Human remains have been found in a small ravine in the inner part of the cave at Mora Cavorso. At least four Radiocarbon dates on human and animal remains dated the Neolithic frequentation of the site to between 6,505 and 6,000 BP (Rolfo et al., 2009, 2012). On Abruzzi slope the earliest human settlements date to the Neolithic period, as evidenced by pottery shards and bits of flint found in the Cola Cave and in the Continenza Cave, where hundreds of human bones were found in the Neolithic levels and were interpreted as burial goods. Similarly, in the Beatrice Cenci Cave and the La Dama Cave, situated above the source of the Liri River, other findings were dated to the Bronze Age (Agostini et al., 1991, Nicolucci, 1877).

In Pre-Roman times the area between Latium and Abruzzi was occupied by several ancient Italic peoples: the Aniene Valley area was inhabited by the Aequi and the Sacco Valley area by the Ernici people. That group belonged to the Osco-Umbrian or Sabellian people, characterized by their use of Osco-Umbrian languages, which were part of the Indo-European language family (Villar, 1997). Conquered by the Romans at the end of the Second Samnite War (326–304 BC), the Aequi and Ernici were subsequently assimilated into Roman culture. In northern Latium many Etruscans settled, the most civilized part of central Italy.

During the all-Roman era, numerous villas were built in the Aniene Valley. Many Roman buildings and constructions, including rustic villas, roads, and plumbing structures, are in the Simbruini mountains area, a sign of a high Roman presence in this area (Giuliani 1979; Troccoli 1985; Tomei 1988; Mari 1995).

Christianity spread rather early throughout its territories, as confirmed by the various shrines and monasteries in the region (Pani Ermini & Giordani, 1978). The valley around Subiaco was also chosen by St. Benedict in the early sixth century AD for the establishment of twelve small monasteries located along the Aniene River (Egidi et al., 1904; Fiore 1994; Appetecchia 2010). From the tenth century onward, the territories were affected by the spread of "Battlement," or an increase in human settlements in internal areas (Marchetti Longhi 1963; Delogu 1978; Belli Barsali 1980; Travaini 1990).

ETRUSCAN PEOPLE: AN UNRESOLVED ENIGMA?

The first case analyzed concerns the use of genetics in the resolving of the Etruscans' origin. The Etruscan culture developed and prospered in the ninth century BCE. In ancient "Etruria," which encompassed Tuscany, northern Latium, and western Umbria (Figure 2). The culture and language of the Etruscan people were distinctively different from those of the Italic populations and, for this reason, two controversial scenarios have been proposed regarding their origin: the Etruscans would have either an Italian origin or a direct descendants form an immigrant group of Aegean/Anatolian ancestry (Beekes 2003). In recent years, several genetic investigations have tried to shed light on the origin of this mysterious people. The first works were focused on ancient DNA and modern DNA belonging to both ancient Etruscans and modern Tuscans, respectively. Through mitochondrial DNA analysis, a genealogical discontinuity was highlighted between the ancient Etruscans and the modern Tuscans, along with a possible gene flow from the Eastern Mediterranean shores (Vernesi et al., 2004, Belle et al., 2006). In 2007 Achilli and colleagues analyzed the maternal genetic background (mitochondrial DNA) of only three villages: Murlo, Casentino, and Volterra. They showed that, in modern Tuscans, a notable number of the maternal lineages (5.3% of the mtDNA haplotypes) were present only in the Anatolian populations. This genetic distribution enabled researchers to conclude that there had been a recent and direct link between Tuscany and the near east (Achilli et al., 2007). These considerations were confirmed in 2009 by the complete mitochondrial DNA analysis, which identified a typical Anatolian lineage in Tuscany (mitochondrial DNA haplogroup U7) and allowed us to estimate, using the molecular clock technique, a historical time frame for the arrival of the Anatolian lineage in Tuscany (1,100 to 2,300 ya B.P.) (Brisighelli et al., 2009). However, several demanding works, based on the DNA of Etruscans and medieval individuals and the mitochondrial DNA of modern Tuscans, confirmed the significant differences between modern Tuscans and Etruscans and suggested that links between the Tuscan and Anatolian populations could be made before the onset of Etruscan culture, during Neolithic times (5000 ya BP). Nevertheless, they sustain that the scarceness of genetic data would make it impossible to conclusively demonstrate an Etruscan origin outside Italy (Ghirotto et al., 2013, Tassi et al., 2013, Guimaraes et al., 2009). Although genetic studies have denied the Anatolian origin of the Etruscan people, this historical enigma could be resolved through analysis of the DNA of the Tuscans. In 2014 many genetic variants (540,000 SNPs) of present-day Tuscans were compared with those of other population groups from Europe and the Middle East. This analysis highlighted the presence of a 25–34% Middle Eastern component in modern Tuscans, confirming that their most probable geographic origin was Eastern Anatolia/Southern Caucasus. This validates the theory proposed by the ancient historian Herodotus on the origin of Etruscans (Pardo-Seco et al., 2014, Gomez-Carballa et al., 2015). It also suggests evident and significant links between recent modern Anatolian and Tuscan peoples. However, because of the vastness and complexity of this historical event, the debate among archaeologists, linguists, and geneticists remains open.

Figure 2. Map of the Etruscan territories.

THE PERMEABLE BARRIER OF LANGUAGE: THE CROATIAN-SPEAKING MINORITY OF MOLISE

A small ethnic community shares not only its language, religion, and patterns of social interaction, but also a common history, which very often becomes a myth. However, the language represents the natural way for closing human groups with each other, influencing social structure and the genetic composition and could act as a barrier, increasing the genetic differentiation between populations. Many migrations to Italy scattered numerous linguistic minorities along the peninsula, including Albanians, Croatians, Franco-Provençals, Greeks, etc. (Salvi, 1975; Toso, 2006, Destro Bisol et al., 2008).

Among the various groups speaking non-Italian languages, the Croatian-speaking community of Molise represents the smallest ethno-linguistic minority (Figure 3). The ancestors of the present inhabitants arrived in this part of Italy in the sixteenth century, from the hinterland of central Dalmatia, and were established in the mountainous inland of Molise. They founded Acquaviva Collecroce, San Felice del Molise, and Montemitro, where Croatian is still spoken today (Salvi, 1975). Croatian has been spoken for about five hundred years and during its long period of contact with the Italian and Molisian dialects, it took new words and phonological features (Sujoldzic, 2004).

From a genetic point of view, these populations were assayed in order to identify admixture insights and understand whether the language was really a barrier. At first, twelve genetic blood markers were analyzed, highlighting a range of genetic admixtures between the Italian and Balkan populations (Biondi et al., 1990). But the deepest analysis was provided by Babalini and colleagues in 2005, when they assayed the maternal genetic background of the Croatian-speaking community of Molise, populations of central Italy, and several Croatian islands. This work indicated that a fraction of the maternal gene pool of the three communities of Molise derives equally from the Croatian parental population and the Italian receptor population of Molise. Moreover the Croatian-speaking community lies genetically very close to other Italian populations, highlighting the high penetration of Italian maternal lineages into their gene pool (Babalini et al., 2005). Overall, the results demonstrated that these communities were not reproductively isolated, despite the influence of social and linguistic components. This condition did not amaze: in fact, over the centuries the female component, which was accounted for in this study, has been influenced more by migration among neighbor populations than the male one. It is plausible that, if the male component (Y-chromosome) were analyzed, it could reveal a higher degree of genetic isolation. However, there were other linguist minorities in Italy that maintained the genetic signs of the native land. In southern Italy, a genetic admixture between Albanian and Italian populations was estimated and confirmed for Calabrian Arbereshe (Albanian-speaking minority of Calabria and Sicily), in accordance with linguistic hypotheses (Sarno et al., 2015). These works demonstrated that, despite the strong cultural identity linked to language, genetics allowed for a more complete account of the real origin of this population.

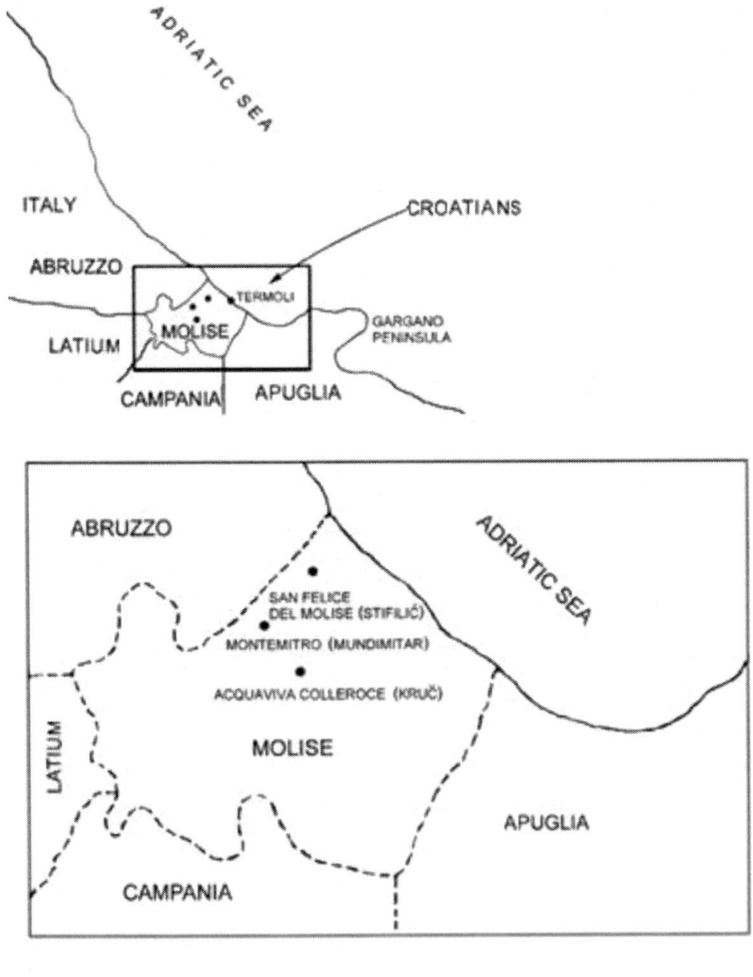

Figure 3. Geographical map showing the location of the villages, where Croatian is still spoken today.

GENETIC INSIGHTS OF FORGOTTEN HISTORICAL EVENTS IN MOUNTAIN COMMUNITIES IN CENTRAL ITALY

In the last case we report studies on an isolated population in central Italy, highlighting how forgotten historical events were discovered through genetics. Single historical events can influence traditions and leave important signatures in the genetic structures of populations (Trupiano, 2012). But what happened? When were these historical events forgotten? Could genetics help to emerge these memories? In 2009 an exploratory genetic analysis on a small mountain village in central Italy (Jenne and Vallepietra, province of Rome) highlighted low frequencies of maternal lineages from the near east (Messina et al., 2010). This study enabled a wide genetic study of all areas in order to understand why there were maternal lineages from the near east/Balkans. It focused on seven mountain communities in central Italy: six villages in Latium (three in the province of Rome: Jenne, Vallepietra, Saracinesco; three in the province of Frosinone: Trevi nel Lazio and Filettino in the Upper Aniene Valley, and Piglio in the Sacco Valley) and one village in Abruzzi (Cappadocia in the province of L'Aquila) (Figure 4).

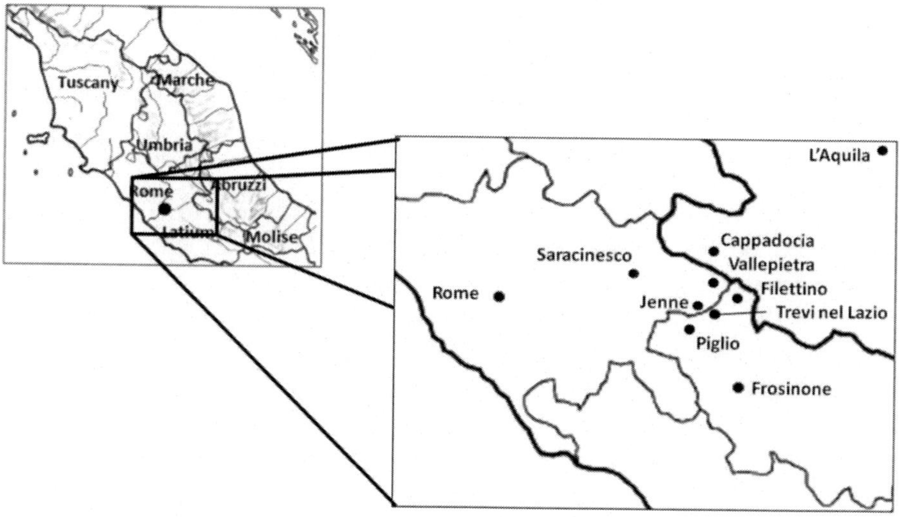

Figure 4. Maps of Italy showing the locations of the villages from Upper Aniene Valley.

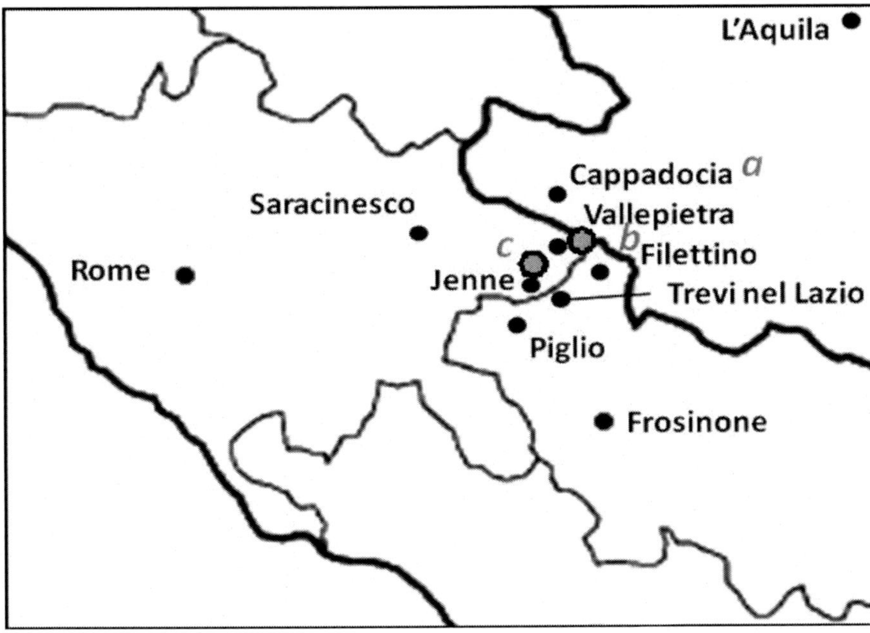

Figure 5. Map of the sampled villages. The grey characters signal interesting locations for study: a) the name of this village coincided with the Anatolian region's name; b) the location of the Santissima Trinità shrine; c) the location of the Saint Blasé Chapel.

These seven communities share features that make them particularly amenable for investigation, namely, they are located in mountain areas (600 to 1,100 meters a.s.l.) and do not belong to a specific ethnic minority. Over the centuries, these villages have maintained their biodemographic stability, and there is historical evidence of early settlements dating back to at least before the Roman conquest of Italy (third century BCE; Gatti, 2008).

The sampling was carried out from 2009 to 2011 by two of us (F. M. and M. F. R.). The maternal analysis was conducted at the Center of Molecular Anthropology for Ancient DNA at the University of Rome "Tor Vergata," directed by O. R., while the paternal analysis was conducted at the Carabinieri Scientific Investigation and Universitad de Barcelona "Pompeu Fabra." Samples were collected from unrelated instances of ascertained maternal and paternal descent. The donors were selected only if their mother, father, and four grandparents had been born in the village. To avoid possible hidden relationships, a form was submitted to each individual sampled, together with informed consent; in the form the donator had to indicate her/his personal data, names, surnames, and the birthplace of four grandparents. 470 individuals were analyzed, of which 237 were males, obtained from seven mountain communities in central Italy. Interestingly, in all seven villages analyzed, lineages not commonly found in Western Europe were found in both sexes. In particular, in Jenne a paternal lineage (Q haplogroup of Y chromosome) from western Asian was detected (Messina et al., 2015). This lineage seems to confirm some kind of correlation between these mountain communities and the people living in western Asia, because a similar link was absent in Italians (Brisighelli et al., 2012), though it was found in 2% of the Turkish population (Cinnioglu et al., 2004) and in nearly 6% of Iranians, Ossets, and Avars (Balanovsky et al., 2011, Grugni et al., 2012). The high frequency of Q haplotypes in Caucasus and the putative origin supported the hypothesis for a small introgression from the Caucasus. These findings are supported by historical sources, suggesting that central Italy could have hosted an Armenian settlement since the beginning of the Common Era (Gaiane` Casnati, 1991). Several toponymic elements scattered throughout the central Apennines seem to further support this suggestion. The name "Cappadocia" one of the sampled villages, could refer to the central Anatolian region, which was formerly a part of the Armenian kingdom in the Middle Ages. Moreover, the surname "Armeni," which means "Armenians" in Italian, is carried by about 5% of the population of Jenne and is documented in notary deeds held at the State Archives of Rome as pertaining to the population of Jenne during the sixteenth and seventeenth centuries (Marini-Dettina, 2011a/b). Another feature was taken into account: the presence of Christian religious elements of Oriental origin. In fact, in Vallepietra the oriental style of the Santissima Trinità shrine was well known by art historians (Rami Ceci, 2003), while there is an ancient chapel (c. fifth century) around Jenne dedicated to Saint Blaise, who was an Armenian Catholic bishop in the third century ACE (Mutafin, 1999) (Figure 5). In conclusion, the central Apennines provides a genetic portrait of past events that could be linked to the Caucasian area, especially for the male structure. However, this data did not allow us to determine when this genetic introgression happened. Rather, it indicated that this isolated area of central Italy was also influenced by migration from far-reaching countries, and that the current people are the result of an admix event.

REFERENCES

Achilli, A., Olivieri, A., Pala, M., Metspalu, E., Fornarino, S., Battaglia, V., Accetturo, M., Kutuev, I., Khusnutdinova, E., Pennarun, E., Cerutti, N.,

Di Gaetano, C., Crobu, F., Palli, D., Matullo, G., Santachiara-Benerecetti, A.S., Cavalli-Sforza, L.L., Semino, O., Villems, R., Bandelt, H.J., Piazza, A. & Torroni, A. (2007). Mitochondrial DNA variation of modern Tuscans supports the near eastern origin of Etruscans. *Am J Hum Genet,* 80, 759-68.

Agostini, S., Coubray, S., De rossi Mazzorin, J., D'ercole, V., Remoti, E. (1991). Indagini preliminari nella Grotta Beatrice Cenci (Cappadocia AQ). In *Bollettino di Archeologia,* 8: 61-71. Rome, Italy.

Appetecchia, A. (2010). *Maestranze e tecniche edilizie nei monasteri sublacensi.* In M. C. Somma, De Re Monastica II: dalla nascita alla morte, 509 Cantieri e maestranze nell'Italia medievale. Spoleto, Italy: Fondazione Centro Italiano di Studi sull'Alto Medioevo.

Babalini, C., Martinez-Labarga, C., Tolk, H.V., Kivisild, T., Giampaolo, R., Tarsi, T., Contini, I., Barac, L., Janicijevic, B., Martinovic Klaric, I., Pericic, M., Sujoldzic, A., Villems, R., Biondi, G., Rudan, P. & Rickards, O. (2005). The population history of the Croatian linguistic minority of Molise (southern Italy): a maternal view. *Eur J Hum Genet,* 13, 902-12.

Balanovsky, O., Dibirova, K., Dybo, A., Mudrak, O., Frolova, S., Pocheshkhova, E., Haber, M., Platt, D., Schurr, T., Haak, W., Kuznetsova, M., Radzhabov, M., Balaganskaya, O., Romanov, A., Zakharova, T., Soria Hernanz, D.F., Zalloua, P., Koshel, S., Ruhlen, M., Renfrew, C., Wells, R.S., Tyler-Smith, C. & Balanovska, E. (2011). Parallel evolution of genes and languages in the Caucasus region. *Mol Biol Evol,* 28, 2905-20.

Beekes, R.S.P. (2003). *The origins of the Etruscans.* Amsterdam, Holand: Nederlandse Akademie van Wetenschappen.

Belle, E.M.S., Ramakrishnan, U., Mountain, J.L. & Barbujani, G. (2006). Serial coalescent simulations suggest a weak genealogical relationship between Etruscans and modern Tuscans. *Proc Natl Acad Sci USA,* 103, 8012-8017.

Belli Barsali, I. & Carbonetti C. (1980). *Lazio medievale.* Ricerca topografica su 33 abitati delle antiche diocesi di Alatri, Anagni, Ferentino, Veroli. Rome, Italy:. Eds. Multimediale.

Biondi, G., Rickards, O., Fuciarelli, M., Purpura, M., Mariani, M., Guidi, A.M., De Vito, G. & Terrenato, L. (1990). Genetic polymorphisms in the Croatian ethno-linguistic minority of Italy. *Gene Geogr,* 4, 71-9.

Biondi, G., Vienna, A., Pena Garcia, J.A. & Mascie-Taylor, C.G. (2005) Isonymy and the structure of the Provencal-Italian ethnic minority. *J Biosoc Sci,* 37, 163-74.

Boattini, A., Luiselli, D., Sazzini, M., Useli, A., Tagarelli, G. & Pettener, D. (2011). Linking Italy and the Balkans. A Y-chromosome perspective from the Arbereshe of Calabria. *Ann Hum Biol,* 38, 59-68.

Boattini, A., Pedrosi, M.E., Luiselli, D. & Pettener, D. (2010). Dissecting a human isolate: Novel sampling criteria for analysis of the genetic structure of the Val di Scalve (Italian Pre-Alps). *Ann Hum Biol,* 37, 604-9.

Brisighelli, F., Blanco-Verea, A., Boschi, I., Garagnani, P., Pascali, V.L., Carracedo, A., Capelli, C. & Salas, A. (2012). Patterns of Y-STR variation in Italy. *Forensic Sci Int Genet,* 6, 834-9.

Brisighelli, F., Capelli, C., Alvarez-Iglesias, V., Onofri, V., Paoli, G., Tofanelli, S., Carracedo, A., Pascali, V.L. & Salas, A. (2009). The Etruscan timeline: a recent Anatolian connection. *Eur J Hum Genet,* 17, 693-6.

Capocasa, M., Anagnostou, P., Bachis, V., Battaggia, C., Bertoncini, S., Biondi, G., Boattini, A., Boschi, I., Brisighelli, F., Calo, C.M., Carta, M., Coia, V., Corrias, L., Crivellaro, F., De Fanti, S., Dominici, V., Ferri, G., Francalacci, P., Franceschi, Z.A., Luiselli, D., Morelli, L., Paoli, G., Rickards, O., Robledo, R., Sanna, D., Sanna, E., Sarno, S., Sineo, L., Taglioli, L., Tagarelli, G., Tofanelli, S., Vona, G., Pettener, D. & Destro Bisol, G. (2014). Linguistic, geographic and genetic isolation: a collaborative study of Italian populations. *J Anthropol Sci,* 92, 201-31.

Casnati, G. (1991). Presenze armene in Italia. Testimonianze storiche ed architettoniche. Venice/Padoa, Italy: AA. VV De Luca Edizioni. *Gli Armeni in Italia.*

Cinnioglu, C., King, R., Kivisild, T., Kalfoglu, E., Atasoy, S., Cavalleri, G.L., Lillie, A.S., Roseman, C.C., Lin, A.A., Prince, K., Oefner, P.J., Shen, P., Semino, O., Cavalli-Sforza, L.L. & Underhill, P.A. (2004). Excavating Y-chromosome haplotype strata in Anatolia. *Hum Genet,* 114, 127-48.

Delogu, P. (1978). Aspetti degli abitati nella regione sublacense. Atti della Società Romana di Storia Patria. 101, 17-34.

Destro Bisol, G., Anagnostou, P., Batini, C., Battaggia, C., Bertoncini, S., Boattini, A., Caciagli, L., Calo, M.C., Capelli, C., Capocasa, M., Castri, L., Ciani, G., Coia, V., Corrias, L., Crivellaro, F., Ghiani, M.E., Luiselli, D., Mela, C., Melis, A., Montano, V., Paoli, G., Sanna, E., Rufo, F., Sazzini, M., Taglioli, L., Tofanelli, S., Useli, A., Vona, G. & Pettener, D. (2008) Italian isolates today: geographic and linguistic factors shaping human biodiversity. *J Anthropol Sci,* 86, 179-88.

Egidi, P., Giovannoni, G., Hermann, G. (1904). *I monasteri di Subiaco,* Roma.

Fiore, M.G. (1994). *Sublaqueum – Subiaco.* Tra Nerone e S. Benedetto. Roma.

Gatti, S. (2008). *Gli Ernici nel quadro delle popolazioni italiche del Lazio.* Rome, Italy: Picuti.

Ghirotto, S., Tassi, F., Fumagalli, E., Colonna, V., Sandionigi, A., Lari, M., Vai, S., Petiti, E., Corti, G., Rizzi, E., De Bellis, G., Caramelli, D. & Barbujani, G. (2013). Origins and evolution of the Etruscans' mtDNA. *PLoS ONE,* 8, e55519.

Giuliani, C. F. (1979). Il territorio tiburtino nell'antichità. *Atti e Memorie della Società Tiburtina di Storia e d'Arte,* 52, 55-63.

Gomez-Carballa, A., Pardo-Seco, J., Amigo, J., Martinon-Torres, F. & Salas, A. (2015). Mitogenomes from The 1000 Genome Project reveal new Near Eastern features in present-day Tuscans. *PLoS ONE,* 10, e0119242.

Grugni, V., Battaglia, V., Hooshiar Kashani, B., Parolo, S., Al-Zahery, N., Achilli, A., Olivieri, A., Gandini, F., Houshmand, M., Sanati, M.H., Torroni, A. & Semino, O. (2012). Ancient migratory events in the Middle East: new clues from the Y-chromosome variation of modern Iranians. *PLoS ONE,* 7, e41252.

Guimaraes, S., Ghirotto, S., Benazzo, A., Milani, L., Lari, M., Pilli, E., Pecchioli, E., Mallegni, F., Lippi, B., Bertoldi, F., Gelichi, S., Casoli, A., Belle, E.M.S., Caramelli, D. & Barbujani, G. (2009). Genealogical Discontinuities among Etruscan, Medieval, and Contemporary Tuscans. *Mol Biol Evol,* 26, 2157-2166.

Jobling, M.A. & Tyler-Smith, C. (2003). The human Y chromosome: an evolutionary marker comes of age. *Nat Rev Genet,* 4, 598-612.

Marchetti Longhi, G. (1963). Il castello di Piglio e i suoi feudatari. *Bollettino dell'Istituto di Storia e di Arte del Lazio Meridionale.* 1, 25-53.

Mari, Z. (1995). La valle dell'Aniene nell'antichità. *Atti e Memorie della Società Tiburtina di Storia e d'Arte.* 68, 25-52.

Marini-Dettina, A. (2011a). *Famiglie di Vallepietra.* Rome, Italy: unpublished.

Marini-Dettina, A. (2011b). Famiglie di Jenne. Rome, Italy: unpublished.

Messina, F., Finocchio, A., Rolfo, M.F., De Angelis, F., Rapone, C., Coletta, M., Martinez-Labarga, C., Biondi, G., Berti, A. & Rickards, O. (2015). Traces of forgotten historical events in mountain communities in Central Italy: A genetic insight. *Am J Hum Biol.* 27, 508-19.

Messina, F., Scorrano, G., Labarga, C.M., Rolfo, M.F. & Rickards, O. (2010) Mitochondrial DNA variation in an isolated area of Central Italy. *Ann Hum Biol,* 37, 385-402.

Mutafian, C. (1999). *Roma-Armenia. Rome,* Italy: De Luca.

Muttoni, G., Scardia, G., Kent, D.V., Swisher, C.C. & Manzi, G. (2009). Pleistocene magnetochronology of early hominin sites at Ceprano and

Fontana Ranuccio, Italy. *Earth and Planetary Science Letters,* 286, 255–268.

Nicolucci, G. (1877). *La grotta Cola presso Petrella di Cappadocia* (third editon). Naples, Italy: Atti dell'Accademia delle Scienze.

Pani-Ermini, L. & Giordani, R. (1978). Note di topografia religiosa della Ciociaria in età paleocristiana e altomedievale: una messa a punto. In AA. VV. *Il paleocristiano in Ciociaria, Atti del Convegno* (Fiuggi 1977). Roma. 78-94.

Pardo-Seco, J., Gomez-Carballa, A., Amigo, J., Martinon-Torres, F. & Salas, A. (2014). A genome-wide study of modern-day Tuscans: revisiting Herodotus's theory on the origin of the Etruscans. *PLoS ONE,* 9, e105920.

Rami Ceci, L. (2003). Sassi e templi. *Il luogo antropologico tra cultura e ambiente*. Rome, Italy: Eds. Armando.

Rolfo, M.F., Salari, L., Zarattini, A.(2009). Nota preliminare sulle indagini archeologiche presso la Grotta Mora di Cavorso (Jenne, Roma). In G. Ghini, Lazio e Sabina 5, *Scoperte Scavi e Ricerch*e (pp.15-22). Rome, Italy: Atti del quinto Incontro di Studi sul Lazio e la Sabina (Roma 2007).

Rolfo, M. F., Lelli, R., Martinez Labarga, C., Passacantando, D., Scorrano, G., Salari, L., Rickards, O. (2012). La comunità neolitica di Grotta Mora Cavorso a Jenne (Roma): osservazioni deposizionali, paleobiologiche e faunistiche. In Negroni Catacchio, N., Atti del Decimo Incontro di Studi. L'Etruria dal Paleolitico al Primo Ferro. *Lo stato delle ricerche* (pp. 131-143). Milan, Italy: Cent. Studi di Preist. e Archeol.

Salvi, S. (1975). *Le lingue tagliate: storia delle minoranze linguistiche in Italia.* Milan, Italy: Rizzoli.

Sarno, S., Tofanelli, S., De Fanti, S., Quagliariello, A., Bortolini, E., Ferri, G., Anagnostou, P., Brisighelli, F., Capelli, C., Tagarelli, G., Sineo, L., Luiselli, D., Boattini, A. & Pettener, D. (2015). Shared language, diverging genetic histories: high-resolution analysis of Y-chromosome variability in Calabrian and Sicilian Arbereshe. *Eur J Hum Genet.*

Soares, P., Achilli, A., Semino, O., Davies, W., Macaulay, V., Bandelt, H.J., Torroni, A. & Richards, M.B. (2010). The archaeogenetics of Europe. *Curr Biol,* 20, R174-83.

Sujoldzic, A. (2004). Vitality and erosion of Molise Croatian dialect. *Coll Antropol,* 28 Suppl 1, 263-74.

Tagarelli, G., Fiorini, S., Piro, A., Luiselli, D., Tagarelli, A. & Pettener, D. (2007). Ethnicity and biodemographic structure in the Arbereshe of the province of Cosenza, southern Italy, in the XIX century. *Coll Antropol,* 31, 331-8.

Tassi, F., Ghirotto, S., Caramelli, D. & Barbujani, G. (2013). Genetic evidence does not support an Etruscan origin in Anatolia. *Am J Physcal Anthr,* 152, 11-8.

Thuiller, J. P. (2008). Gli Etruschi. *La prima civiltà italiana.* Turin, Italy: Eds. Lindau.

Toso, F. (2006). Lingue d'Europa. *La plularità linguistica dei Paesi europei fra passato e presente.* Milan, Italy: Saggi Baldini and Castoldi, Dalai.

Tomei, M. A. (1988). *Il suburbium di Roma in età imperiale: forme di insediamento e proprietà della terra in alcune aree lungo l'Aniene e la Via Tiburtina.* Atti e Memorie della Società tiburtina di Storia e d'Arte, 61, 57-130.

Travaini, L. (1990). *Due castelli medievali: Affile ed Arcinazzo Romano.* Roma.

Troccoli, G. (1985). La valle dell'Aniene in età romana. *Quaderni del Centro di studio per l'archeologia etrusco-italica,* 7, 175-7.

Trupiano, V. (2012). The Patrimonialization of DNA. An Italian Case Study: Genetic Park of Cilento and Vallo di Diano. *Int Politic Anthr* 5, 135–156.

Vernesi, C., Caramelli, D., Dupanloup, I., Bertorelle, G., Lari, M., Cappellini, E., Moggi-Cecchi, J., Chiarelli, B., Castri, L., Casoli, A., Mallegni, F., Lalueza-Fox, C. & Barbujani, G. (2004). The Etruscans: a population-genetic study. *Am J Hum Genet,* 74, 694-704.

Villar, F.(1997). *Gli Indoeuropei e le origini dell' Europa.* Bologna, Italy: Il Mulino.

In: Italy
Editor: Gina Pittman

ISBN: 978-1-63484-408-6

Chapter 2

LAND TAKE AND URBAN RENEWAL: A STUDY CONCERNING SARDINIAN URBAN AREAS (ITALY)

Corrado Zoppi*[*], *Sabrina Lai and Anania Mereu
Dipartimento di Ingegneria Civile, Ambientale e Architettura,
Università di Cagliari, Cagliari, Italy

ABSTRACT

Land take is the "Change of the amount of agriculture, forest and other semi-natural and natural land taken by urban and other artificial land development" (European Environment Agency, 2013).

In 2011, the European Commission (EC) put in evidence that an important milestone for the EU should be to reach the goal of no net land take by 2050, and to take under strict control the impact on land-taking processes of the EU policies in the new Structural Funds programming period (2014-2020) (Communication of the EC to the European Parliament COM (2011) 571 of 20.9.2011).

In this paper we address two intertwined issues in the conceptual context of land take. First, we analyze and compare the characteristics of a fundamental phenomenon related to land take, that is the change from non-artificial to urbanized areas in medium-sized towns. We study the phenomenon in some coastal cities of Sardinia and try to put in evidence

[*] Corresponding author: Tel.: Italy–(0)70–6755216, telefax: Italy–(0)70–6755215, E-mail: zoppi@unica.it.

a few lessons, both in terms of determinants of urbanization of non-artificial land and of planning policy implications.

Second, we consider one of the most important land take-related issues, that is urban renewal concerning historic centers' preservation. Indeed, restoration and reuse-based policies of historic centers is one of the most effective ways of preventing land-taking processes, especially in terms of increasing the housing stock.

Under this perspective, we analyze the determinants of the perceived quality of the urban environment of Cagliari's historic district through a survey addressed to the resident families and administered through a questionnaire. The survey is interpreted through a discrete choice model whose results put in evidence several important inferences related to residential satisfaction. These inferences could be very helpful for the definition of future public policies for a substantial improvement in the quality of life of historic centres, related to public services and infrastructure endowment, for example public parking, refuse collection and retail shops.

1. INTRODUCTION

The EC indicates that land take in the EU amounted to more than 1,000 km^2 per year between 1990 and 2000, decreasing to about 920 km^2 between 2000 and 2006 (European Commission, 2011), and that, as a consequence, the objective of no net land take by 2050 would imply a decrease rate of about 800 km^2 per year.

Land take in Italy parallels the difficult general situation of other EU countries. Figures at the national level put in evidence that in 2009 a 7.3% of the Italian land had an artificial land cover (European Commission, EUROSTAT, 2010), with an average growth rate of about 6% between 1990 and 2000 and of about 3% between 2000 and 2006 (ISPRA, 2011, p. 479). The implementation of analyses of land-taking processes at the regional level is problematic since currently available geographic databases and information systems do not provide systemic information on the phenomenon (CRCS, 2012).

However, some Italian regional administrations, such as Lombardy and Sardinia, have set up regional information systems that may help address land-taking processes. The geographic information systems of these regions allow to estimate land take and to relate it with spatial, economic and planning-policy related variables, and to infer on correlations between such variables and the land-taking phenomenon.

We study the land-taking process through the land cover maps of Sardinia, made available in 2003 and 2008 by the Sardinian regional administration.[1] The results and inferences of our study could be easily generalized to other Italian and EU regions, under the necessary condition that geographic databases and maps were available for these contexts as well.

Moreover, we consider urban renewal concerning historic centers' preservation, as one of the most important questions concerning mitigation of land taking processes. We see urban renewal as an effective comprehensive approach to prevent land take, since it increases housing supply.

In the framework of regional and urban planning processes of Sardinia, in the context of the Regional landscape plan (RLP), established by the Decision of the Sardinian Regional Government (DSRG) no. 36/7 of 5 September 2006, the Implementation plans of historic centers (IPHCs) are planning tools which implement the Planning implementation code (PIC) of the RLP into the so-called "Areas characterized by historic settlements." For these areas, the PIC defines a set of prescriptive rules and planning criteria (articles nos. 51-53 of the part of the PIC related to "Cultural and historic spatial framework," which is defined in articles nos. 47-59). More precisely, article no. 52 identifies the IPHC as a plan which has to be necessarily approved through the cooperation of the Sardinian regional administration and a municipality as a necessary precondition for a municipality to exert its ruling power over the local transformation processes related to the municipal spatial jurisdiction, which implies a considerable pressure on local administrators in order to implement valuable and effective planning processes concerning municipal historic centers.

As a consequence, in the planning processes of the IPHCs, heavily influenced by the control of the technical staff of the regional offices, a strong consistency and implied uniformity do show up as: i. a strong attention to historical, typological and morphological characteristics in terms of the territorial analysis of historic urban settlement systems, which are identified by the RLP as "Centers of antique and primary development"; ii. a strong

[1] The 1:25,000 "Land Use Map of the Region of Sardinia – 2003 Edition" and "New Land Use Map of the Region of Sardinia - 2008 Edition" are actually two land cover maps that cover the whole island. Data were obtained mainly from photo-interpretation of aerial photographs and satellite images, together with other vector data sets and on-site surveys. The maps' minimum mapping unit (Longley et al., 2001, 151) equals 0.5 ha in urban areas and 0.75 ha in rural areas. Both maps can be freely downloaded from: http://www.sardegnageoportale.it/index.php?xsl=1598&s=141401&v=2&c=8831&t=1 [accessed: October 2014].

prescriptive ruling framework characterized by a markedly conservative attitude.

It is therefore not surprising, given this context, that the RLP, the first statutory landscape plan with regional dimensions produced in Italy under the new legislation, focused on the coastal zone because of the complexity of development conflicts arising from tourism (on which a large part of the economy of the island relies) and other development, and owing to the fact that thirteen out of the fourteen previous landscape plans covering coastal areas, which contained some restrictions on coastal development, had been quashed between 1998 and 2003.

Following approval of the plan in 2006, restrictions and prohibitions (on development of land and on certain changes in land uses) stemming from the plan are currently in force, in order to protect a part of the island considered economically strategic and environmentally sensitive. Restrictions and prohibitions are set out by the plan by means of a system of rules.

The planning activity of the regional administration of Sardinia has undergone a deep change after the approval of the RLP, which establishes the directions for nearly any future planning activity in Sardinia, and requires that actual sectoral and local plans, as well as plans for protected areas, be changed to comply with its directions.

This paper is organized as follows. In the second section we propose the definition of land take for the purpose of this paper and an analysis of land-taking processes of Sardinia (Italy). We discuss the influence of the factors/variables found relevant on land take that could be taken into account to define regional planning policies to limit or possibly prevent land take, and, by doing so, help implementing the EC recommendation on no net land take by 2050 into EU regional policies.

In the third section, we analyze the determinants of the perceived quality of the urban environment of the historic center of Cagliari, the regional capital city of Sardinia, through a survey addressed to resident households and administered through a questionnaire. The survey is interpreted through a discrete choice model whose results put in evidence several important inferences related to residential satisfaction. These inferences could be very helpful for the definition of future public policies for the substantial improvement of the quality of life within historic centers, related to public services and infrastructure endowment, for example public parking, refuse collection and retail shops.

In the concluding section, we put in evidence implications of the outcomes of the interpretive models implemented in the two previous sections in terms of urban and regional policies, and directions for future research.

2. A Perspective on Sardinian Land Take

As we put in evidence above, the EC considers to reach no net land take by 2050 as an important milestone for a roadmap to a resource-efficient Europe. One of the most dangerous consequences of land take is soil sealing, but other related phenomena are soil contamination and erosion, decrease in soil organic content and in agricultural production and productivity. In a recent study published by the Italian Institute of Urban and Regional Planning (CRCS, 2012), a systematic discussion on the impacts of land-taking processes is proposed; such impacts are grouped as follows: i. impacts on the carbon cycle: a decline in the power of the soil's organic content to fix carbon dioxide in the atmosphere and an increase in concentration of carbon di-oxide generated by the mineralization of the carbon present in the excavated soil of new urban developments; ii. impacts on the water cycle and microclimate (soil sealing implies: a significant decline in stored ground- and underground water; an increased flood risk due to the rising quantity of rainfalls which run directly into rivers, augmenting their levels, turbulence, and sediments in the water; and, impacts on urban microclimate, since the decrease in soil evapotranspiration power is likely to generate an increase in atmospheric temperature); iii. impacts on biodiversity: land-taking processes cause the soil's impoverishment and, as a consequence, the loss of huge quantities of microorganisms, which could mitigate soil contamination, filter percolation waters and make nutrients available for vegetation and pastures; iv. impacts on agricultural production: potential agricultural crops are heavily and progressively hindered by land take and soil sealing.

However, from the relevant paragraph concerning land–taking processes of the EC communication quoted above (paragraph 4.6), it is rather difficult to derive a rigorous definition of land take, which should be based on its unwanted impacts in order to effectively address and mitigate their consequences.

Let us consider, for example, the Land Use and Cover Areas frame Survey (LUCAS) of EUROSTAT (European Commission, EUROSTAT, 2010), and the COoRdination de l'INformation sur l'Environnement (CORINE) Land Cover vector map (CLC) of the European Environment Agency (EEA) of the EU (European Environment Agency, 2013). In LUCAS, "artificial land," that is land taken by land-taking processes, is classified into two main groups, that is "built-up" and "non built-up" areas, where the former are further classified according to the number of floors of their buildings, while a separated sub-group is represented by greenhouses (Technical reference, document C-3 - Land use and Land Cover: Nomenclature, pp. 14-16). In CLC, "artificial surfaces" are classified into four groups (CORINE Land cover - Part 2: Nomenclature, p. 1): i. urban fabric; ii. industrial, commercial and transport units; iii. mine, dump and construction sites; and, iv. artificial, non-agricultural vegetated areas. Even though both LUCAS and CLC address the issue of artificial land cover, propose definitions of artificial vs. non-artificial land cover, and identify artificial and non-artificial areas, they greatly differ from each other.

The above example shows that it is quite difficult and controversial to frame and identify a precise measure of land take, which in some way can make it difficult to implement rigorous quantitative studies on this subject. From this perspective, there are at least two relevant general issues to be taken into account. First, it is rather controversial to state univocally that land take is always negative in terms of the negative impacts above mentioned, since there are types of land take which do not generate those impacts. For example, soil sealing, one of the most dangerous impacts, is not a necessary consequence of land-taking processes, as indicated by the EC, which puts in evidence that soil sealing is limited to about a 50% of the land taken: "In the EU, more than 1,000 km^2 are subject to 'land take' every year for housing, industry, roads or recreational purposes. About half of this surface is actually 'sealed'." (EC COM(2011) 571, paragraph 4.6)

Second, there is the trade-off critique. This critique considers land take as a process caused by strong pressure in favor of settlement development, which implies that the land taken will increase its market value once new land uses displace existing uses. So, why, in principle, should existing uses be preferred over the new ones? Moreover, is a prohibitionist, normative, approach the most efficient way to prevent the negative impacts of land-taking processes from taking place in the long run? Neo-liberist positions support this critique (see, for example: MacCallum, 2003; Moroni, 2007). From this point of view, heavy taxation on land rent could possibly be the most effective means to counter demand for land take.

Table 1. Variables relevant to explain land-taking processes and their descriptive statistics. Data are collected at municipal level

Variable	Definition	Mean	S.D.
LAND TAKE	2003-2008 percent change from non-artificial to artificial land cover (sources: Corine Land Cover Map of Sardinia – 2003 Edition, next "CLCMS03," level 1; Corine Land Cover Map of Sardinia – 2008 Edition, next "CLCMS08," level 1)	0.53	0.99
PARCSIZE	Average size of areas classified as non-artificial in 2003 and artificial in 2008 (ha) (sources: CLCMS03, level 1; CLCMS08, level 1)	0.33	0.29
ACCESS	Endowment of roads connecting regional town and city centers per unit of municipal land area (km/km^2) (source: Spatial Dataset of the Regional Geographic Information System of Sardinia, next SDRGISS[2])	0.95	0.47
DISTCAPC	Distance of a municipality from the regional capital city, Cagliari (km) (source: Google Maps)	126.45	71.17
DISTNEAC	Distance from the closest province administrative center (km) (source: Google Maps)	30.98	16.67
DISCOAST	Weighted average distance of areas classified as non-artificial in 2003 and artificial in 2008 from the shoreline (km); weight = area size (sources: CLCMS03, CLCMS08, SDRGISS)	21.02	13.99
CONSAREA	Total protected area: parks, reserves, etc. (ha) (sources: CLCMS03, CLCMS08, SDRGISS)	1,342.74	2,632.62
NATAR	Landscape components with an environmental value, defined as natural and seminatural areas that change from non-artificial to artificial land cover in 2003-2008 (ha) (sources: CLCMS03, CLCMS08, RLP spatial dataset)	11.67	26.05
AGRFORAR	Landscape components with an environmental value, defined as agricultural and forestry areas that change from non-artificial to artificial land cover in 2003-2008 (ha) (sources: CLCMS03, CLCMS08, SDRGISS, RLP spatial dataset)	25.70	50.83
COASTRIP	Percentage of a municipality's area included in the coastal strip (sources: SDRGISS, RLP spatial dataset)	11.18	24.96
OLDPLAN	Area classified under the planning code in force before 2006 as area where land transformations and new developments are almost totally forbidden that changes from non-artificial to artificial land cover in 2003-2008 (ha) (sources: CLCMS03, CLCMS08, SDRGISS)	14.85	43.06
DENSITY	Population density in 2003 (residents per square kilometer) (source: web site Sardegna Statistiche: http://www.sardegnastatistiche.it [accessed: May 2015])	76.55	209.39

[2] Available from the Regional Geoportal, at: http://www.sardegnageoportale.it/index.html [accessed: October 2014].

In this paper we do not propose ethic narratives or value judgments on land take, but we analyze land-taking processes in order to detect which factors, and possibly to what extent, can be considered relevant to explain the phenomenon.

We implement our analysis with reference to the Sardinian region. Sardinian is located to the west of Central Italy, off the west coast just south to the French island of Corsica. Sardinia has detailed land-cover maps based on the CLC classification, available for 2003 and 2008, that make it possible to analyze the dynamic of land cover through the comparison of land cover classes consistent with each other. Therefore, we use the CLC-based maps of Sardinia to study land take processes, since the LUCAS data, available for 2008 only, would have not allowed us to study land take as a dynamic process.

In the CLC classification, non-artificial surfaces are classified into four classes (at Level 1): i. agricultural areas; ii. forests and semi-natural areas; iii. wetlands; and, iv. waterbodies. The land-taking process is identified in this study as the shift of areas from non artificial classes in 2003 to the artificial land-cover class in 2008. Sardinia has experienced an increase in artificial land from 2.75% in 2003 (66,206 hectares) to 3.22% in 2008 (77,516 hectares).

Table 1 shows the variables we consider relevant to explain land-taking processes, and their descriptive statistics. The variables refer to spatial units identified with the 377 municipalities of Sardinia.

An analytic and detailed discussion on the reasons of the choice of this set of variables as relevant to explain land-taking processes and on the estimate of the econometric model which uses these factors as covariates of land take is proposed in a recent essay by the authors (Zoppi and Lai, 2014).

We regard the Sardinian land-taking process as related to factors identified as relevant variables in several studies concerning land take in the mainstream literature, through censored and OLS regression models. We tentatively consider a set of variables which includes location-related and physical determinants, planning code rules, and socio-economic factors.

We find that there is a double agglomeration effect, since land-taking processes are positively and significantly related to high population density and high concentration of land which changes its status from non-artificial to artificial. This indicates that saving non-artificial land, or limiting land take, could be effectively supported by low-density settlements and extensive and light land-taking processes, since the concentration of land take in a limited number of municipalities would imply a larger extent of land which becomes artificial, being non-artificial in the first place.

Secondly, the more a municipality is accessible, the more it is suitable to land-taking processes, which indicates that balancing accessibility opportunities would be a strategic regional policy in order to limit the concentration of land take and, ultimately, to mitigate the agglomeration effect which characterizes land take. This goal could be reached by increasing the endowment of public roads connecting regional town and city centers to small municipalities, giving particular care to road connections to the regional capital and province cities. Moreover, the lower a municipality's proximity to the regional administrative capital city (Cagliari) or to the nearest province administrative center (DISTCAPC, DISTNEAC), the less its suitability to land take, which is an argument in favor of balancing accessibility as well.

Thirdly, we find that the presence and size of protected areas is negatively and significantly connected to land take, as expected. So, conservation of natural resources, habitats and environment could be strategically important in order to deal with land-taking processes, and to influence their territorial layout. This is also confirmed by the estimates related to the covariate OLDPLAN, which is positively correlated to the change of land from non-artificial to artificial. This indicates that the more conservative planning rules are weakened, the more land-taking processes occur, which is what happened (in 2003) in the areas where the old regional landscape plans were not in force any more. A similar phenomenon is put in evidence by the covariates AGRFORAR and NATAR, which are positively, even though significantly only in the case of the Tobit model, correlated to PERLTAKE, which suggests, as before, that the more conservative planning rules are weakened, the more land-taking processes occur: in the case of AGRFORAR and NATAR it is evident that the conservation character of the RLP PIC was weak if non-artificial areas defined as landscape components with an environmental value were allowed to change their status from non-artificial to artificial land between 2003 and 2008.

The fact that protection of nature, environment and natural resources matters is also put in evidence by the absence of correlation between land-taking processes and the variables COASTRIP and DISCOAST, which indicates that land take was not a coastal phenomenon in the period 2003-2008. Since in the Eighties and in the Nineties the Sardinian regional land-taking processes were almost exclusively concentrated in coastal municipalities, the non-coastal characterization of land take between 2003 and 2008 could only be related to the conservative planning rules that the regional landscape plans in force before 2006 and the RLP, from year 2006 on, have implemented.

3. A Historic Center's Life Quality-Related Survey

Planning and historic centers' preservation can be considered synergetic concepts (Frank and Petersen, 2002) and activities. Indeed, effectiveness of preservation is founded on analytic and detailed planning which should implement policies aimed at preserving the characteristics of historic urban tissues and at boosting their capability of being attractive for the most important contemporary urban functions, such as housing, retail shops, and financial, insurance and real estate firms (Mueller et al., 2005). Moreover, it could be important to maintain the sense of identity of the people who live there. Hence, it is necessary to implement participatory processes involving resident families, entrepreneurs, employees and shoppers (Comune di Reggio Emilia, 2005, 2011).

One of the main components of sustainable planning approaches to the preservation of historic centers implies the availability of detailed and, as much as possible, complete information on residents' perceived needs and expectations. In this paper, we implement a methodological approach to collect this information, which can be used as an important reference point for the definition of planning policies for the historic centers' preservation (Fantin, 2013).

We use the econometric approach based on Discrete Choice Models (DCM) (Ben-Akiva and Lerman, 1985; Greene, 1993; Greene and Hensher, 2010) to identify and analyze the residents' needs and expectations concerning the spatial organization of the historic neighborhoods where they live, which provides inferences on residential satisfaction's determinants (Lu, 1999). In this framework, determinants are grouped into three distinct categories as follows: i. structural characteristics of the residential unit; ii. neighborhood's characteristics; iii. respondents' social and demographic characteristics.

The DCM-based analysis is implemented through a questionnaire delivered to the residents of the Cagliari's (Sardinia, Italy) historic center's neighborhoods (Comune di Cagliari, 2011). Through their participation in the experiment, resident families' representatives should increase their awareness of the most important issues related to the spatial organization of their urban living areas, therefore allowing us to explore correlations between perceived urban quality and the three types of determinants mentioned above.

Qualitative and quantitative inferences on the correlations between residents' level of satisfaction and its determinants, generated by the DCM-based revealed preferences approach, imply important arguments and indications on planning policies related to the spatial organization of the historic centers, which will be discussed in the final part of the paper, starting from the results of the DCM-based case study concerning the historic center of Cagliari. Among the most important planning policy-related issues which can be effectively addressed through the information coming from the analysis of the revealed preferences and needs of residents are the following: availability of car parks in the historic centers, refuse collection policy, location of public services and infrastructure.

The model's implementation implies a definition of residential satisfaction and residential environment. According to Galster (1987), residential satisfaction entails an assessment of the qualitative difference between the present and the perceived best conditions of a house's and of a neighborhood's characteristics. The quality of a residential environment depends on the house's and neighborhood's characteristics. Social aspects that account for relations with neighbors are particularly relevant (Amerigo and Aragones, 1997). Therefore, we propose the variables listed in Table 2 as possible determinants of residential satisfaction, and we use them to implement our DCM-based estimates. We use an *Ordered Logit* model, where the dependent variable is the residential satisfaction related to neighborhood characteristics (Williams, 2005).

The model's results show that respondents older than 40 seem to express a higher satisfaction degree than younger people, which can be related to expectations' and needs' changes during lifetime. The results concerning the variable D_AGE's ratio indicate that if the respondent is older than 40, the probability of having a higher satisfaction degree increases by a 1.85 factor. With reference to the gender-related variable D_GEN, women express a level of satisfaction higher than men's, which can be explained through comparatively higher emotional connection with the residential location-related relationships. As expected, the presence of relatives in the neighborhood has a positive influence.

The variables related to the tranquillity of the respondent's house location and to the quality of relationships with neighbors, D_QUIET and D_RELATION, show positive odds ratios, 2.06 and 4.84 respectively, as expected. Respondents with a low educational level reveal a lower satisfaction degree than graduates. Practitioners show a satisfaction level lower than employees, whereas students, unemployed and retired people exhibit a higher satisfaction than employees.

Table 2. Historic center's life quality-related variables and their descriptive statistics

Variable	Definition	Mean	St.Dv
NEIG_SAT	Ordered categorical variable of residential satisfaction related to neighborhood that can take five values, as follows: 1-low level of satisfaction; 2-low-medium; 3-medium; 4-medium-high; 5-high	3.77	1.06
D_AGE	Dummy variable that represents the age of the respondent (1 if the age is less than 40, 0 otherwise)	0.13	0.33
D_GEN	Dummy variable that represents the gender of the respondent (1 for females and 0 for males)	0.63	0.48
D_QUIET	Dummy variable that states the perceived tranquillity of the place where the respondent lives (1 if it is a calm place, 0 otherwise)	0.71	0.46
D_RELAT	Dummy variable that shows the presence of respondent's resident relatives in the historic center (1 if there are relatives, 0 otherwise)	0.35	0.48
D_RELATION	Dummy variable related to the goodness level of the relationships between the respondent and his neighbors (1 if good or excellent, 0 otherwise)	0.77	0.42
EDUC_LEV	Set of dummy variables related to the respondent's educational qualification (EDUC_LEV1 is 1 if the respondent has a university degree or a diploma, 0 otherwise; EDUC_LEV2 equals 1 if the respondent has a primary or high school certificate, 0 otherwise)	0.14	0.35
JOB	Set of dummy variables related to the respondent's job (JOB1 is 1 if the respondent is a public employee, 0 otherwise; JOB2 is 1 if the respondent is a practitioner, 0 otherwise; JOB3 is 1 if the respondent is a student or an unemployed person, a retiree or a housewife, 0 otherwise)	0.13	0.34
D_SCHOOL	Dummy variable related to the presence of schools or university facilities near the respondent's house	0.65	0.48
D_MARKET	Dummy variable related to the presence of a market shop near the respondent's house	0.65	0.48
D_POST	Dummy variable related to the presence of a post office near the respondent's house	0.71	0.46
D_PHARM	Dummy variable related to the presence of a pharmacy near the respondent's house	0.80	0.40
D_GARD	Dummy variable related to the presence of a garden near the respondent's house	0.45	0.50
D_BUSST	Dummy variable related to the presence of bus stops near the respondent's house	0.79	0.41
D_PARK	Dummy variable related to the absence of parking spaces in the neighborhood of the respondent's house	0.83	0.38

Variable	Definition	Mean	St.Dv
D_STRLIGHT	Dummy variable related to the presence of a deficient street lighting in the neighborhood of the respondent's house	0.14	0.35
D_CRIME	Dummy variable related to the presence of criminal activity in the neighborhood of the respondent's house	0.16	0.36
D_REFUS	Dummy variable related to the presence of problems of refuse collection in the neighborhood of the respondent's house	0.44	0.50
D_TRAFF	Dummy variable that represents the presence of traffic in the neighborhood of the respondent's house	0.36	0.48
D_WATLOS	Dummy variable related to the presence of water losses in the neighborhood of the respondent's house	0.10	0.30
D_NOISE	Dummy variable related to the presence of noise in the neighborhood of the respondent's house	0.42	0.49
D_PAV	Dummy variable related to the presence of incoherent paving in the neighborhood of the respondent's house	0.07	0.26
D_HOUSAT	Dummy variable related to the respondent's satisfaction concerning the house where he/she lives (1 if the satisfaction degree is medium-high or high, 0 otherwise)	0.84	0.36
DUR_RES	Ratio between the number of years the respondent has been living in the house and his/her age (percentage of lifetime the respondent has spent in the house)	0.57	0.33

Variables related to the presence of schools, post offices and bus stops have a positive influence on the satisfaction related to the neighborhood of the respondent's house, as expected. In particular, variable D_BUSST shows an odds ratio of 2.81.

The variable related to the presence of markets and gardens seem to have a negative effect, which is not consistent with expectations. The presence of pharmacies does not seem to have any effects on the level of satisfaction related to the neighborhood where the respondent's house is located.

The absence of parking spaces, the lack of sufficient street lighting, the presence of crime in the neighborhood and problems connected to refuse collection, that is variables D_PARK, D_STRLIGHT, D_CRIME and D_REFUS, exhibit negative odds ratios. The traffic-related variable D_TRAFF has a negative influence and it shows an odds ratio nearly equal to 0.57.

The variable related to water losses, D_WATLOS, has a positive effect, and it exhibits an odds ratio nearly equal to 2.30. Noise-related and paving-related variables, that are D_NOISE and D_PAV, show negative effects. Moreover, the variable related to satisfaction concerning the house where the respondent lives reveals a high odds ratio, nearly equal to 2.59. Finally, the variable related to the number of years the respondent has been living in the house exhibits, contrary to expectations, a low odds ratio, nearly equal to 0.74.

CONCLUSION

We tentatively consider a set of variables which includes location-related and physical determinants, planning code rules, and socio-economic factors. As we stated in the second section, we do not assume ethic narratives or value judgments on land take. Nevertheless, the findings imply a set of policy statements which can be taken into account in order to influence land-taking processes. Agglomeration effect (both in terms of land which becomes artificial being non artificial in the first place, and of residential concentration) increases the intensity of land take. As a consequence, extensive urbanization and planning codes which prevent the artificialization of vast contiguous areas should be effective in saving-up non-artificial land. A balanced accessibility of regional cities and towns and a comprehensive regional policy concerning protection of nature, natural resources, environment and endangered species and habitats should be important as well. Moreover, supporting restrictive planning rules concerning new development in the coastal strip helps to counter and limit land take.

The outcomes of the DCM-based model entail important implications for future planning policies. The restricted availability of public services and infrastructure is a very influential factor in determining low satisfaction. As a consequence, to provide the historic center with important services such as bus stops is highly likely to be effective. Thus, an important policy proposal could be to improve the public transportation network in order to make it easier for the residents to commute from the historic center to other urban destinations. With reference to traffic problems, which is another negative factor, it seems that policies that entail an increase in parking space very close to the historic center boundary could generate a positive effect on residential satisfaction related to neighborhood, which may possibly imply the pedestrianization of the most part of Cagliari's historic district as well. As far as relationships with neighbors are concerned, the outcomes of the model's implementation show that good relationships have a positive effect on residential satisfaction. Hence, it is reasonable to propose a policy that contributes to improve relationships between neighbors, such as, for instance, making community centers available for residents.

Since more liveable houses and neighborhoods endowed with high-quality public services and infrastructure make peripheral and outbound locations more attractive than the historic centers', a key element to rebalance the historic centers' displacement is to increase the quality of inner cities' houses and neighborhoods. As an example, in the case of Cagliari, the model's implementation shows that attractiveness of the historic center could improve substantially if the availability of post offices, bus stops and retail shops increased. Moreover, the model's outcomes indicate that a comprehensive planning approach to public transportation, parking, and pedestrian paths would help making the historic center comparatively more attractive, since these issues are much more important for the historic center than for other urban neighborhoods.

This paper analyzes the Sardinian land-taking process as related to factors which are identified as relevant variables concerning land take in several studies of the mainstream literature. The methodology can be easily replicated and exported with reference to other Italian and European contexts and results could be straightforwardly comparable. Policy implications of the findings could be a point of reference for future Italian and European land-use and planning policies which entail a careful consideration of the negative impacts of artificialization of land, as the Communication of the EC to the European Parliament COM(2011) 571 of 20.9.2011 puts in evidence. Future research should also relate to the construction of a spatial data infrastructure to monitor and control land-taking processes, with a view to the objective of no net land take by 2050.

The results obtained with reference to Cagliari's historic center allow generalization for two reasons. On the one hand, no similar empirical studies have been implemented to analyze the determinants of residential satisfaction related to neighborhood in other Italian conurbations by means of a DCM-based approach. On the other hand, it is not possible to compare the situation of the urban area of Cagliari to a situation in which a more flexible, participatory, faster and bottom-up planning process was implemented. This kind of situation would have probably encouraged people to lobby in favor of effective planning policies concerning the historic center, since the established planning process has been developed quite homogeneously in all of Italy, and counter-examples are very rare.

References

Amerigo, M. & Aragones, J. I. (1997). A theoretical and methodological approach to the study of residential satisfaction. *Journal of Environmental Psychology*, *17* (1), 47–57.

Ben-Akiva, M. & Lerman, S. R. (1985). *Discrete Choice Analysis: Theory and application to travel demand*. Cambridge, MA, United States: The MIT Press.

Comune di Cagliari [Municipality of Cagliari]. (2011). *Piano particolareggiato del Centro storico [Implementation plan of the historic center]*, available at http://www.comune.cagliari.it/portale/it/ppcs.page; jsessionid=8BF4671D5DE9617D3293821977AB3879 [accessed: May 2015].

Comune di Reggio Emilia [Municipality of Reggio Emilia] (2005). *Piano strategico per la valorizzazione della città storica di Reggio Emilia [Strategic Plan for the Increase of the Value of the Historic City of Reggio Emilia]*, available at http://www.municipio.re.it/download/cittaStorica/doc_indirizzi_piano_strategico.pdf [accessed: May 2015].

Comune di Reggio Emilia [Municipality of Reggio Emilia]. (2011). *Piano strutturale comunale di Reggio Emilia, P3.2 Strategie e azioni per la Città storica [Municipal Structural Plan for the Reggio Emilia, p.3.2 Strategies and Actions for the Historic City]*, available at http://www.municipio.re.it/Sottositi/PSCRE.nsf/0/5EA2768B7578B3D9C12575A5003B1AA0?opendocument&FT=P [accessed: May 2015], and at http://www.municipio.re.it/download/pscre/1PSC/Elaborati_tecnici/P3.2_Strategie_e_azioni_per_la_citta_storica.pdf [accessed: May 2015].

CRCS (Centro di Ricerca sui Consumi di Suolo [Research Center for Land-taking Processes]) (2012). *Rapporto 2012 [2012 Report]*. Rome, Italy: INU Edizioni.

European Commission. (2011). *Report on Best Practices for Limiting Soil Sealing and Mitigating its Effects. 2011*, available at http://ec.europa.eu/environment/archives/soil/pdf/sealing/Soil%20sealing%20-%20Final%20Report.pdf [accessed: May 2015].

European Commission, EUROSTAT. (2010). *Land Cover/Use Statistics (LUCAS)*, available at http://epp.eurostat.ec.europa.eu/portal/page/portal/lucas/introduction [accessed: May 2015].

European Environment Agency. (2013). *CORINE Land Cover*, available at: http://www.eea.europa.eu/publications/COR0-landcover [accessed: May 2015].

Fantin, M. (2013). Il Masterplan del Centro storico di Vicenza [The Masterplan of the historic center of Vicenza]. *Urbanistica, 55 (150-151)*, Supplement, 1-34.

Frank, K. & Petersen, P. (2002). *Historic Preservation in the USA*. Berlin, Germany: Springer.

Galster, G. C. (1987). Residential segregation and interracial economic disparities: A simultaneous-equations approach. *Journal of Urban Economics*, *21* (1), 22-44.

Greene, W. H. & Henser, D. A. (2010). *Modeling Ordered Choices: A Primer.* New York, NY, United States:Cambridge University Press, New York.

Greene, W. H. (1993). *Econometric Analysis*. New York, NY, United States: Prentice Hall.

ISPRA. (2011). *Annuario dei dati ambientali 2011 – Tematiche in primo piano [Yearbook of 2011 Environmental Data – Mainstream Themes]*, available at http://www.isprambiente.gov.it/files/pubblicazioni/stato ambiente/tematiche2011 [accessed: May 2015].

Longley, P. A., Goodchild, M. F., Maguire, D. J. & Rhind, D. W. (2001). *Geographic Information. Systems and Science*. Chichester, United Kingdom: John Wiley & Sons.

Lu, M. (1999). Determinants of residential satisfaction: ordered logit vs. regression models. *Growth and Change*, *30* (2), 264-287.

MacCallum, S. H. (2003). The enterprise of community: market competition, land, and environment. *Journal of Libertarian Studies*, *23* (4), 1-15.

Moroni, S. (2007). *La città del liberalismo attivo [The City of Active Liberalism]*. Milan, Italy: Città Studi.

Mueller, B., Curwell, S. & Turner, J. (2005). Un modello per il miglioramento delle LUDA: lo sviluppo del collaborative strategic goal oriented programming [A model for improving the LUDA: the implementation of the collaborative strategic goal oriented programming]. *Urbanistica Dossier*, *8 (74)*, 14-19.

Williams, R. A. (2005). *Gologit2: A Program for Generalized Logistic Regression/Partial Proportional Odds Models for Ordinal Dependent Variables*, available at http://www.stata.com/meeting/4nasug/gologit2.pdf [accessed: May 2015].

Zoppi, C. & Lai, S. (2014). Land-taking processes: An interpretive study concerning and Italian region. *Land Use Policy*, *36* (1), 369-380.

In: Italy
Editor: Gina Pittman

ISBN: 978-1-63484-408-6

Chapter 3

SEASONAL ANALYSIS OF CROP EVAPOTRANSPIRATION BASED ON PENMAN-MONTEITH EQUATION IN PO RIVER DELTA, ITALY

Mărgărit-Mircea Nistor*
University of Modena and Reggio Emilia,
Department of Chemical and Geological Sciences, Modena, Italy

ABSTRACT

The recent global warming has negative impact on water balance, natural systems, and agriculture activities. Assessing the crop evapotranspiration, we estimated the fraction of water consumed by vegetation. We have calculated the crop evapotranspiration in Po River Delta with Penman-Monteith equation using daily input data and through Budyko method the water balance was calculated. In relationship with crop growth, we defined five stages for 2013: initial, crop development, mid-season, late season, and the cold season. The field coefficient was calculated to estimate the evapotranspiration for entire Po River Delta. In the mid-season was observed that field coefficient has a major value of 1.341 and crop evapotranspiration values than exceed 290 mm/month were identified in summer period: 291.37 mm in June, 332.93 mm in July, and 295.19 mm in August. The major rate of potential

* E-mail address: renddel@yahoo.com.

evapotranspiration was carried out for July, with a value of 248.77 mm. With 149 mm of precipitation in summer, the first results indicate that Po River Delta had a water deficit in mid-season. We consider that these findings represent a quantitative analysis about evapotranspiration in the most important wetland area from Italy and could be useful for hydrologists and climatologists.

Keywords: seasonal crop evapotranspiration, Penman-Monteith, Po River Delta

INTRODUCTION

Evapotranspiration is an important parameter for water balance. Globally, the surfaces water and aquifers recharge depend directly by precipitations, temperatures, and effects of land cover on evapotranspiration phenomena. In the last seventy years, in many places of the world was proposed and computed the potential evapotranspiration (ET_0) based on different meteorological data. Early, Thornthwaite bring award a evapotranspiration formula, that taken into account the monthly temperature data and specific latitude radiation parameter (Thornthwaite, 1948). In the same year, Penman (1948) porposed a formula for evaporation of open water, more complex due to the enrichment of the parameters. Monteith (1965, 1985) developed a method to computing the water evaporation from vegetations. Later, because of insuficiently definition of aerodynamics and wind function of bulks from the soil in the Penman equation, over time the scientists have tried to adjust the formula. Beven (1979), Allen et al. (1989), Jensen et al. (1990) used a combination between Penman-Monteith methods in theirs papers. In the report of Food and Agriculture Organization of the United Nations (FAO-56), Allen et al. (1998) derived the Penman-Monteith equation integrating the bulk surfaces resistence and function of wind speed.

In the present context of climate change, IPCC (2001) called increases in temperature trend by 0.6°C during the 20 century. The climate warming continue (Shaver et al. 2000; Dong et al. 2013) in current century and for 2100 year the climate projections shows rising in temperature by 3.5°C (Stocks et al. 1998; The Canadian Centre for Climate Modelling and Analysis, 2014) and reduction of precipitation. These results could be affects seriously the water resources and natural systems (Johnes, 1996; Dragoni, 1998; Parmesan and Yohe, 2003; Gianniou and Antonopoulos, 2007; IPCC, 2007). For these reasons, we conducted the present research that proposes an estimation of evapotranspiration based on Penman-Monteith equation. We chose the Po River Delta because is an important wetland in temperate zone, positioned on the 45th parallel north, with a complex vegetation system. Thus, we want to see the crop evapotranspiration (ET_c), considering that crops may absorb a significant quantity of water in wetlands, even if missing the precipitations.

In the past, several authors adopted Penman-Monteith method for estimation of land evapotranspiration and water studies (Rana and Katerji, 1998; Allen, 2000; Abdelhadi, 2000; Sumner and Jacobs, 2005; Gong et al. 2006). Valiantzas (2013) made a standardization of the Penman-Monteith equation that could be used in cases of weather data limitation. Choudhury (1997) determinate the Global pattern evaporation using Penman-Monteith method and satellite data.

GIS techniques were used in the present study in ET_c mapping and area calculations of each category of crop. ArcGIS from ESRI Software have been widely applied in hydrology and earth sciences by McKinnney and Cai (2002), Kattaa et al. (2010), Elfarrak et al. (2014), Costache (2014), Mitof and Prăvălie (2014), Nistor (2014), Nistor and Porumb-Ghiurco (2015). For the estuarine ecosystem analysis and environment indicators assessment were used remote sensing and GIS by Yang (2005). Draper et al. (2003) used GIS applications for management area of Botanical Garden from Lisbon.

The aim of paper is to calculate the ET_c in the study area for 2013 year, based on Penman-Monteith equation. The study provide related data of ET_c values for different stages of crop growth and a one map with distribution of ET_c in identified stages.

Materials and Methods

Study Area

Po River Delta is located in eastern part of Po Plain, at latitude of 45° North (Figure 1). This delta represents the largest wetland from Italy and occupies approximately 568.61 km^2 and continues to expand by 0.6 km^2/year. Together with lakes and lagoons has a 680.01 km^2 in surfaces. Po Delta is formed by six river braches, at contact of Po River flows with Adriatic Sea: Po di Levante, Po di Maistra, Po della Pila, Po delle Tolle, Po di Gnocca, and Po di Goro (Grillotti Di Giacomo, 2008).

The vegetation that growing in this area is generally hydrophilic formed by herbaceous plants and mesophilic trees. The most important forest from Po River Delta is "Gran Bosco della Mesola" recognised as a natural reservation.

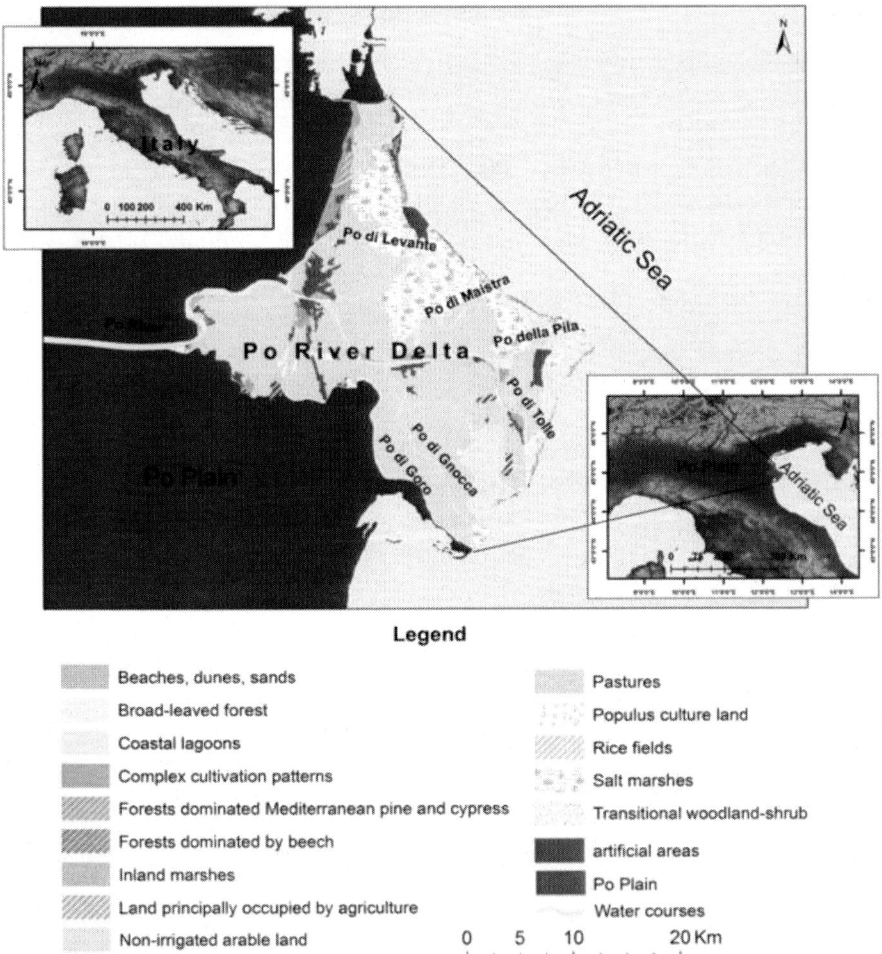

Figure 1. Location of the Po River Delta on the Italy map.

Materials

The materials used in this paper are classified into meteorological data and spatial territory data. Daily data of mean temperature, minimum temperature, maximum temperature, precipitation, mean relative humidity, minimum relative humidity, maximum relative humidity, wind speed, and integral radiation, referred to 2013 year, were used to calculate the ET_c in Po River Delta. The temperature, humidity, and precipitation data belong to Copparo meteorological station, the integral radiation belongs to Ferrara station and the wind speed belong to Colorno station. All meteorological data are courtesy of Regional Agency of Prevention and Environment Emilia-Romagna (ARPA ER). During 2013, in November 15 the Copparo station doesn't working and for another fourteen days was insufficient data registered. For this reason the lost data of relative mean, minimum and maximum humidity were replaced with average data of respective month. This simple calculation not affect our final result, because data lost were in January and December, a period when only the forest and populus culture land area are available to evapotranspiration.

Corine Land Cover data, detailed at 4th level from 2006, and DEM of Europe were used in the spatial location and identifications of main crops in the Po River Delta. These data were integrated in ArcGIS 10.1 from ESRI software, subsequently used for areas calculations and ET_c mapping for intense evapotranspiration season.

Methods

Determination of ET_c Using Penman-Monteith Method

The evapotranspiration computed with Penman-Monteith method becomes a complex matter because of number of variables and for frequency of meteorological data required in equation. Penman-Monteith equation (Eq.(1)) taken into account daily mean temperature, minimum temperature, maximum temperature, precipitation, mean relative humidity, minimum relative humidity, maximum relative humidity, wind speed, integral radiation, and derived variables, to obtain the ET_0. The crop contribution for evapotranspiration phenomena is contained by K_c coefficient (K_c), and multiplied this coefficient by ET_0, the ET_c was carried out (Eq. (8)). The methodology of ET_c calculation with Penman-Monteith equation is presented by Allen et al. (1998) in FAO-56 Paper. Due to complexity of interferences between crops stages and life behaviour of plants in Po Delta is very complicated to analyse the interval time and ET_c of each category of land. In this study has been simplified the strategy at micro scale, in aim to evaluate the seasonality at entire delta scale. For 2013 was chose four stages from

March to October, and the fifth is considered the cold stage: in November and winter month's period. Daily data of ET_c was further transformed in monthly ET_c in aim to mapping the seasonality of evapotranspiration.

$$ET_0 = \frac{0.408\Delta(R_n - G) + \gamma\frac{900}{T+273}u_2(e_s - e_a)}{\Delta + \gamma(1 + 0.34u_2)} \quad \text{(Eq. 1)}$$

where:

ET_0	reference potential evapotranspiration [mm day^{-1}]
R_n	net radiation at the crop surface [MJ m^{-2} day^{-1}]
G	soil heat flux density [MJ m^{-2} day^{-1}], could be ignored because is small in comparison with R_n (Allen et al. 1998)
T	mean daily air temperature at 2 m height [°C]
u_2	wind speed at 2 m height [m s^{-1}]
e_s	saturation vapour pressure [kPa] (Eq.(2))
e_a	actual vapour pressure [kPa] (Eq.(3))
e_s-e_a	saturation vapour pressure deficit [kPa]
Δ	slope vapour pressure curve [kPa°C^{-1}] (Eq.(4))
γ	psychrometric constant [kPa°C^{-1}] (Eq.(5))

$$e_s = \frac{e^{\circ}(T_{max}) + e^{\circ}(T_{min})}{2} \quad \text{(Eq. 2)}$$

where:

e (T_{min})	saturation vapour pressure at daily minimum temperature [kPa]
e (T_{max})	saturation vapour pressure at daily maximum temperature [kPa]

$$e_a = \frac{e^{\circ}(T_{min})\frac{RH_{max}}{100} + e^{\circ}(T_{max})\frac{RH_{min}}{100}}{2} \quad \text{(Eq. 3)}$$

where:

e (T_{min})	saturation vapour pressure at daily minimum temperature [kPa]
e (T_{max})	saturation vapour pressure at daily maximum temperature [kPa]
RH_{max}	maximum relative humidity [%]
RH_{min}	minimum relative humidity [%]

$$\Delta = \frac{4098\left[0.6108 \times exp\left(\frac{17.27T}{T+237.3}\right)\right]}{(T+237.3)^2} \quad \text{(Eq. 4)}$$

$$\gamma = 0.00163 \left(\frac{P}{\lambda}\right) \quad \text{(Eq. 5)}$$

where:

P atmospheric pressure [kPa]
λ latent heat of vaporization [MJ kg^{-1}]

$$P = 101.3 \left(\frac{293 - 0.0065\,z}{293}\right)^{5.26} \quad \text{(Eq. 6)}$$

where:

z elevation above sea level [m]

$$\lambda = 2.501 - (2.361 \times 10^{-3})T \quad \text{(Eq. 7)}$$

$$ET_c = ET_0 K_c \quad \text{(Eq. 8)}$$

where:

ET_c crop evapotranspiration [mm day^{-1}]
ET_0 potential evapotranspiration [mm day^{-1}]
K_c crop coefficient [dimensionless]

The Seasonal Single Crop Coefficient K_c

Each type of plant has the evapotranspiration capacity that is different during the crop growth and especially, in the temperate zone, during the seasons. In the Po River Delta, some of crops are not active in the cold periods; some of them reduced the intensity of evapotranspiration, because of small heat index. Certainly, for cultivated crops and leaved plants are different periods of development. Allen et al. (1998) suggest four crop growth stages for various types of crop: (i) initial, (ii) crop development, (iii) mid-season, (iv) late season. In case of coniferous or evergreen crops, the mid-season stage is continuous.

In Po River Delta, the presence of agriculture land, forest, and transitional woodland-shrubs are predominantly. Forest areas is divided into broad-leaved forest like oak *(Quercus robur)* and beech *(Fagus sylvatica)* and the evergreen forest contains trees like mediterranean pine (*Pinus pinea)* and cypress

(Cupressus sempervirens). The agriculture land is covered by pastures and arable land. In a small part are present the rice fields. For all land types and crops has been assigned a K_c (Table 1) for initial, development, mid-season, and late season stages. For the January, February, November, and December months was considered a reduced K_c, in attempt to obtain a suitable value for cold period. This reason is coming because the crops and small vegetation suspended the activity in winter as consequences of lack of energy. Thus, only the forests and populus culture land resulted active in Po River Delta in cold season. The K_c was assigned after consulted values from FAO-56 Paper, and an average of height was attributed for crops.

Table 1. Seasonal K_c and crop height

Type of land cover/Season	Initial		Crop development		Mid-season		Late season		Cold season	
	K_c	Height (m)	K_c	Height (m)	K_c	Height (m)	K_c	Height (m)	K_c	Height (m)
Land principally occupied by agriculture	1.1	0.2	1.2	0.4	1.35	1.2	1.25	1.2	-	-
Non-irrigated arable land	1.1	0.2	1.2	0.4	1.35	1.2	1.25	1.2	-	-
Complex cultivation patterns	1.1	0.15	1.2	0.3	1.35	1.2	1.25	1.2	-	-
Rice fields	1.05	0.2	1.12	0.5	1.2	1	0.75	1	-	-
Pastures	0.4	0.1	0.6	0.2	0.9	0.25	0.85	0.15	-	-
Transitional woodland-shrub	0.8	3	0.9	3	1	3	0.95	3	-	-
Populus culture land	1.2	4	1.3	4	1.5	4	1.25	4	0.5	4
Forests dominated by beech	1.3	30	1.45	30	1.6	30	1.5	30	0.5	30
Broad-leaved forest	1.3	30	1.45	30	1.6	30	1.5	30	0.6	30
Forests dominated by mediterranean pine and cypress	0.95	10	1	10	1	10	1	10	1	10

To analyze the entire delta ET_c based on unique K_c, we proceed at $K_{c\ field}$ calculation, that contains in formula the height, area, and K_c of crop (Eq.(9)). For each season was obtained a $K_{c\ field}$, used further in the annual trend graphic.

$$K_{c\ field} = \frac{A_{crop\ 1}H_{crop\ 1}K_{c\ crop\ 1}+\cdots+A_{crop\ n}H_{crop\ n}K_{c\ crop\ n}}{A_{crop\ 1}H_{crop\ 1}+\cdots+A_{crop\ n}} \quad \text{(Eq. 9)}$$

where:

A_{crop} field area covered by crop [km²]
H_{crop} crop height [m]
$K_{c\ crop}$ crop coefficient [dimensionless]

Actual Crop Evapotranspiration (AET_c)

Due to elevated temperature in Po River Delta the ET_0 and ET_c register in summer period high values. For this motive, the annual balance of precipitation and crop evapotranspiration in some days is negative. In reality, the crops in irrigated areas or wetlands could produce evapotranspiration even if the rainfall doesn't exist. Using Budyko equation (Eq. (10)) we calculated the AET_c, integrating the precipitation data. Budyko method is significant for water balance study and indicates if the energy is enough to evaporate the precipitation (Gerrits et al., 2009).

$$\frac{AET_c}{P_a} = [(\varphi tan\frac{1}{\varphi})(1 - exp^{-\varphi})]^{0.5} \quad \text{(Eq. 10)}$$

where:

AET_c actual crop evapotranspiration
P_a total annual precipitation
φ aridity index (Eq.(7))

$$\varphi = \frac{ET_c}{P_a} \quad \text{(Eq. 11)}$$

RESULTS

The following findings were identified from the seasonality analysis: ET_0 and ET_c values, ET_c mapping for the five crop stages in 2013, and a water balance was calculated. Figure 2 show the daily trend of ET_0 and ET_c during the crop seasonality in entire Po River Delta. The maximum value of daily ET_0 was calculated in June 21 and had a value of 9.932 mm, and minimum ET_0 was 0.115 mm in January 2. The ET_c register maximum and minimum values in the same days, with 13.316 mm respective 0.086 mm. From the graphic is observed that in cold season the ET_c is less than ET_0, due to subunit $K_{c\ field}$ in this period.

Interestingly is that cumulative ET_0 and ET_c per month reach maximum values in July, 248.77 mm respective 332.93 mm, due to elevated temperature in more days than in June.

Regarding at seasonality of ET_c in Po River Delta, during March to October is concentrated the all crops vital activity. In the initial stage, the ET_c had relative small daily values between 0.259 mm and 4.944 m. Usually, this is the shortest stage of plants, but the numbers of days included in initial stages may vary depending of crop and climate (Allen et al. 1998). In this stage the ground cover is approximately 10% and evapotranspiration is related at soil evaporation (Allen et al. 1998).

In the crop development stage, in Po River Delta was observe a high range of ET_c values, from 1.42 mm/day to 8.967 mm/day. These data reflect the instability of meteorological data and development of the vegetation. In this stage the ground cover starts from 10% and continues to fully cover and includes the flowering and plants growth in height.

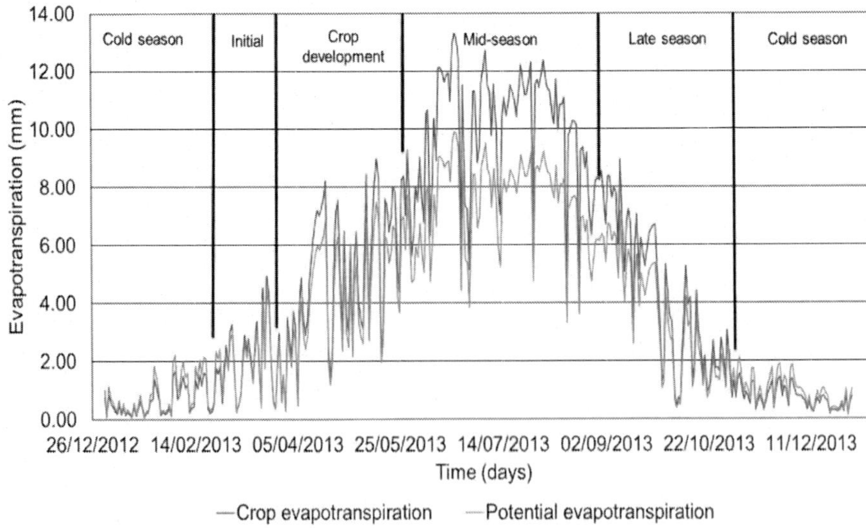

Figure 2. Statistics of daily ET_0 and ET_c in Po River Delta in 2013.

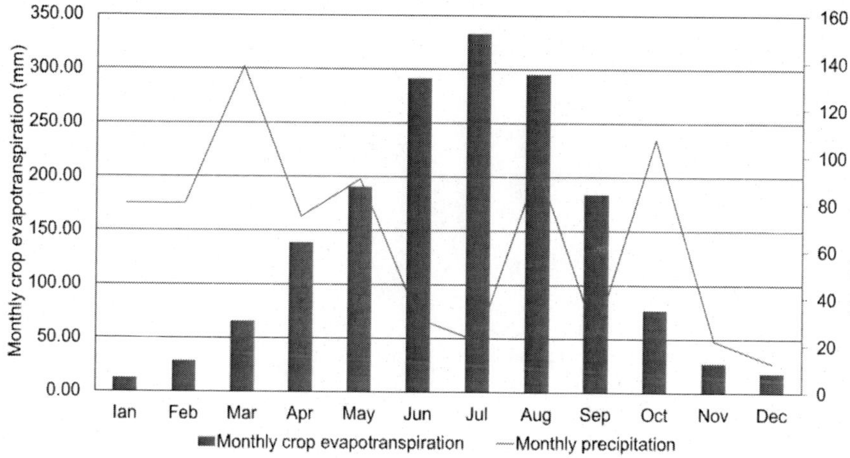

Figure 3. Statistics of monthly ET_c and precipitation in Po River Delta in 2013.

The mid-season is the most complex stage due to intensively aspects of crops. The ground cover becomes fully covered; the maturity of plants is perceptible from yellowing leaves or fruit ripening. The daily ET_c includes values from 4.422 mm to 13.316 mm in Po River Delta and is the highest daily values from the inventoried year.

The late season correspond to harvest or full senescence of plant, when the leaf dropped (Allen et al. 1998). In this period the ET_c values fall gradually from 8.932 mm/day to 0.458 mm/day. This stage and development stage could register high ET_c values only if the crops are irrigated.

In the cold season, the ET_c values in Po River Delta are below 2 mm/day due to low temperature and a small $K_{c\ field}$.

The obtained monthly values of ET_c from May to September exceeded the monthly precipitation. These results are problematic in the real assessment of ET_c. Because is a wetland, Po River Delta has the capacity to evapotranspiration especially in proximity of marshlands. Regarding to forest areas the deficit of water is clear, because of high capacity of evapotranspiration. In the months in which the precipitation was less (Figure 3) than evapotranspiration it was obtained: in June the ET_c score was 291.37 mm, in July ET_c was 332.93 mm, in August ET_c register 295.19 mm, and in September ET_c was 183.73 mm. The spatial distribution of seasonality ET_c is represented in Figure 4.

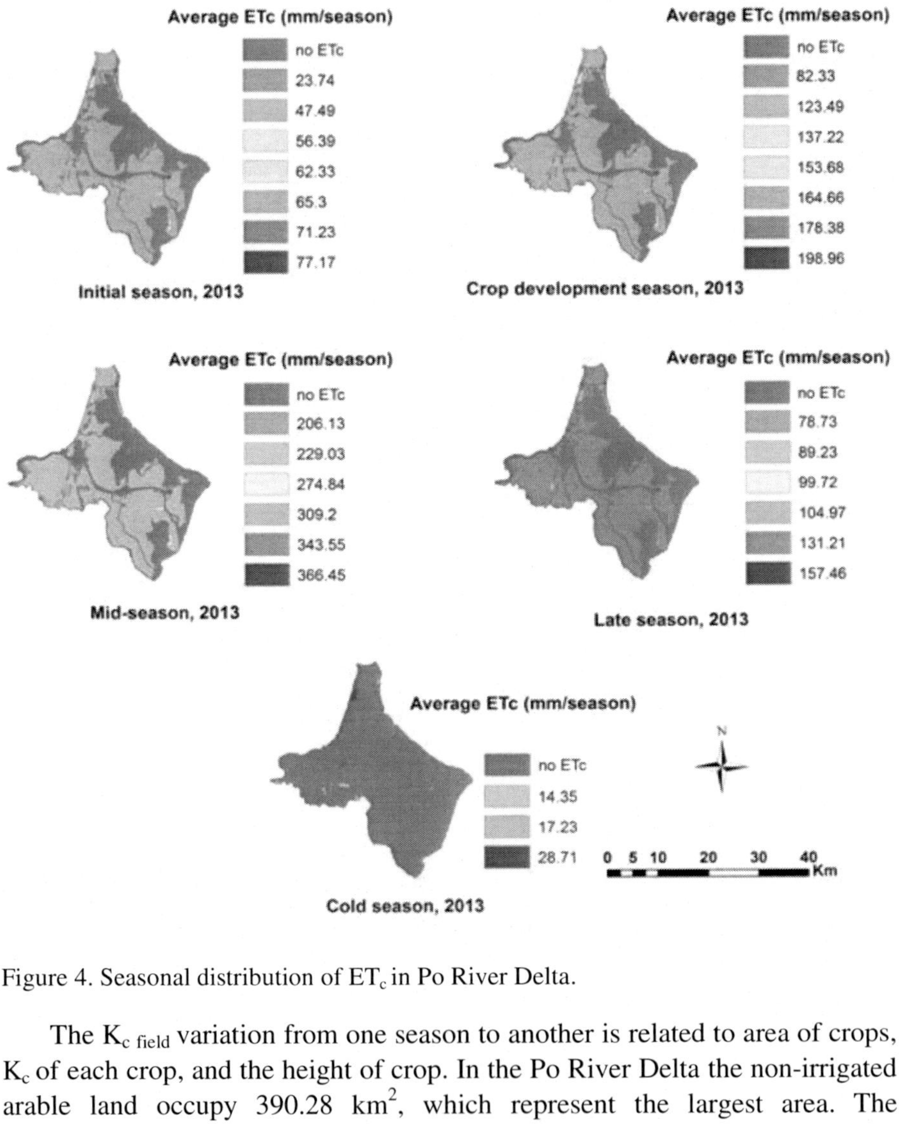

Figure 4. Seasonal distribution of ET_c in Po River Delta.

The $K_{c\ field}$ variation from one season to another is related to area of crops, K_c of each crop, and the height of crop. In the Po River Delta the non-irrigated arable land occupy 390.28 km^2, which represent the largest area. The following are complex cultivation patterns and populous culture land with 17.29 km^2, respective 4.62 km^2. The Mediterranean pine and cypress forest extends on 4.41 km^2. The rice fields occupied 2.85 km^2, while the smallest areas are covered by beech forest and pastures, with 0.49 km^2, respective 0.3 km^2. Considering the above parameters a $K_{c\ field}$ of 1.107 was carried out for initial season, 1.202 for crop development season, 1.341 for mid-season, and 1.242 for late season. In the cold season the $K_{c\ field}$ was calculated combining

K_c of forests and populus culture land, leaving aside other crops. Thus, for winter period was calculated a value of 0.748 for $K_{c\ field}$ in the cold season (Table 2). This value is acceptable for the study area, considering that in frozen period in north China was found a value of 0.4 (Allen et al. 1998).

Table 2. Specific $K_{c\ field}$ values

Season	Initial	Crop development	Mid-season	Late season	Cold season
$K_{c\ field}$	1.107	1.202	1.341	1.242	0.748

Table 3. Annual water balance in Po River Delta using Budyko formula

Annual ETc (mm) calculated with Kc field, after Penman-Monteith equation	1663.398
Aridity index, after Budyko formula	2.139
Precipitation (mm)	777.800
AETc (mm)	705.637
Water surplus (mm)	72.163

Applying the Budyko equation, the aridity index of 2.139 was found for the Po River Delta and the AET_c was calculated: 705.637 mm/year. Making the differences between annual precipitation and AET_c a value of 72.163 mm of non-evapotranspiration water was resulted (Table 3). This findings would seem that almost all rainfall fallen on dry land is consumed by plants and the Po River Delta represent a territory with high ET_c.

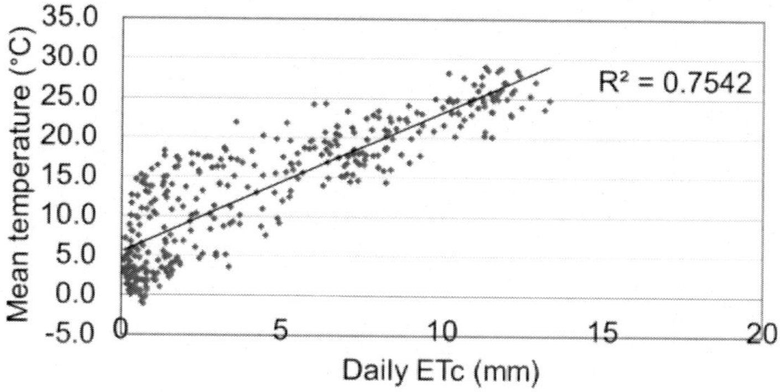

Figure 5. Scatter plot of daily ET_c and mean temperature data.

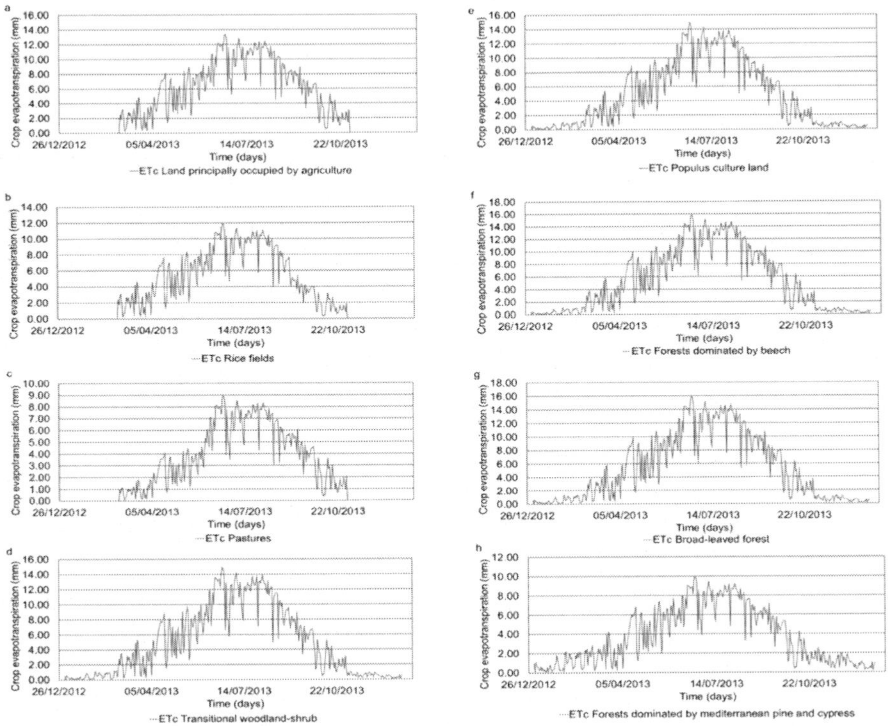

Figure 6. Daily variations of the different crops in Po River Delta. (a) agriculture, (b) rice, (c) pasture, (d) transitional woodland-shrub, (e) populus culture land, (f) beech forest, (g) broad leaved-forest, (h) mediterranean pine and cypress.

Table 4. Annual values of ETc , Aridity index, and AETc related types crops in Po river Delta

Type of crop	ET_c (mm/year)	Aridity index	AET_c (mm)
Land principally occupied by agriculture	1584.63	2.04	698.25
Forests dominated by beech	1946.79	2.50	726.41
Rice fields	1351.67	1.74	670.79
Pastures	985.25	1.27	601.85
Populus culture land	1778.49	2.29	715.06
Transitional woodland-shrub	1181.03	1.52	643.60
Broad-leaved forest	1958.28	2.52	727.10
Forests dominated by mediterranean pine and cypress	1342.71	1.73	669.53

Discussion

The main goal of present paper was to estimate the seasonality ET_c in Po River Delta based on Penman-Monteith equation and mapping the related ET_c in each season for 2013. The overall direction of results indicates the high ET_c during mid-season and late season, in lands covered by forest and agricultural area. The most important factors that influenced the ET_c in Po River Delta are land type and climate conditions. It was observed that in cold and initial seasons the ET_c are less and in other seasons increased due to major of temperature and crop activity. The relationship between ET_c and mean temperature is showed in Figure 5. In the same time we expected to obtain high values for ET_c in mid-season and late season period, but we didn't expect to have a very high deficit in summer, but is possible due to lack of precipitation.

Analyse the daily ET_c for each type of crop, it was observe that beech and other broad-leaved forests register the highly values. Thus, the daily ET_c in June reach at 15.71 mm, while the pastures provide the smallest ET_c, the maximum value calculated was 8.84 mm. The trend is for mediterranean pine and cypress, that not exceed 10 mm in mid-season and has small amplitude of ET_c due to activity in cold season. Figure 6 give the annual trend of main crops land from Po River Delta and Table 4 offers the annual ET_c and AET_c data for each category of crop.

The limitation of study is given by field measurements with evaporation pan or lysimeter. So, the paper contain an empirical approach, with standard data of K_c and crop average height, which could be submit some errors due to field conditions, time gap between seasons or variability of climate.

Conclusion

The ET_0 and ET_c using Penman-Monteith equation were calculated in different stages for 2013, in Po River Delta. Based on results of this study, the following conclusions may be drawn:

- The trends of each category of crop were presented in all set stages.
- Daily and monthly ET_0 and ET_c were computed and spatial ET_c distribution was mapped.

- High values of ET_c were register in mid-season period and ET_c is highest in areas covered by broad-leaved forest.
- Low values of precipitation and elevated temperature in summer produced a water deficit due to high crop water demand.
- During the research was related the AET_c used Budiko method, and then we are able to provide the water balance for 2013.

As a general conclusion, the survey carried out provides simply basis of evapotranspiration in Po River Delta for further studies dealing in biology, hydrology, and climatology. The present study contributes to the research literature on ET_c in Po River Delta, because no work has yet been published on the ET_c in this important wetland area.

ACKNOWLEDGMENTS

The author thanks his aunt Nelia Ciaca for all support during the university studies.

REFERENCES

Abdelhadi, A.W., Takeshi, Hata T., Tanakamaru, H., Tada, A., and M.A. Tariq. 2000. Estimation of crop water requirements in arid region using Penman-Monteith equation with derived crop coefficients: a case study on Acala cotton in Sudan Gezira irrigated scheme. *Agricultural Water Management* 45:203–214.

Allen, R.G., Jensen, M.E., Wright, J.L., and R.D. Burman. 1989. Operational estimates of evapotranspiration. *Agron. J.* 81:650–662.

Allen, R.G., Pereira, L.S., Raes, D., and M. Smith. 1998. Crop evapotranspiration: guidelines for computing crop water requirements. FAO Irrigation and drainage paper 56, Rome.

Allen, Richard G. 2000. Using the FAO-56 dual crop coefficient method over an irrigated region as part of an evapotranspiration intercomparison study. *Journal of Hydrology* 229:27–41.

ARPA ER (Agenzia regionale prevenzione e ambiente Emilia-Romagna), 2011. Climatological data for 2013 at Copparo, Ferrara, and Colorno. Accessed December 17, 2014.

Beven, Keith. 1979. A sensitivity analysis of the Penman-Monteith actual evapotranspiration estimates. *Journal of Hydrology* 44:169–190.

Choudhury, Bhaskar J. 1997. Global Pattern of Potential Evaporation Calculated from the Penman-Monteith Equation Using Satellite and Assimilated Data. *Remote Sens. Environ.* 61:64–81.

Costache, Romulus. 2014. Using GIS techniques for assessing lag time and concentration time in small river basins. Case study: Pecineaga River Basin, Romania. *Geographia Technica* 09(1):31–38.

Grillotti Di Giacomo, M.G. 2008. Atlante Tematico delle Acque d'Italia. Brigati, ISBN 9788887822540, Genova.

Dong, P., Wang, C., and J. Ding. 2013. Estimating glacier volume loss used remotely sensed images, digital elevation data, and GIS modelling. *International Journal of Remote Sensing* 34(24):8881–8892.

Dragoni, Walter. 1998. "Some considerations on climatic changes, water resources and Water needs in the Italian region south of the 43_N." In *Water, Environment and Society in Times of Climatic Change*, edited by Kluwer N. Brown and A. Issar, 241–271, ISBN: 0-7923-5282-3.

Draper, D., Rosselló-Graell, A., Garcia, C., Gomes, C.T., and C. Sérgio. 2003. Application of GIS in plant conservation programmes in Portugal. *Biological Conservation* 113:337–349.

Elfarrak, H., Hakdaoui, M., and A. Fikri. 2014. Development of Vulnerability through the DRASTIC Method and Geographic Information System (GIS) (Case Groundwater of Berrchid), Morocco. *Journal of Geographic Information System* 6:45–58.

Gianniou, S.K., Antonopoulos, V.Z., 2007. Evaporation and energy budget in Lake Vegoritis, Greece. *Journal of Hydrology* 345:212– 223.

Gong, L., Xu C., Chen, D., Halldin, S., and Y.D. Chen. 2006. Sensitivity of the Penman–Monteith reference evapotranspiration to key climatic variables in the Changjiang (Yangtze River) basin. *Journal of Hydrology* 329:620–629.

IPCC, 2001. "Contribution of Working Group I to the Third Assessment Report of the Intergovernmental Panel on Climate Change," In *Climate Change 2001: The Scientific Basis* edited by Houghton, J.T., Ding, Y., Griggs, D.J., Noguer, M., van der Linden, P.J., Dai, X., Maskell, K., and K.A. Johnson, 1-83, Cambridge University Press, Cambridge, United Kingdom and New York, NY, USA.

IPCC, 2007. "Contribution of Working Group I to the Fourth Assessment Report of the IPCC," In *Climate Change 2007: The Physical Science Basis* edited by Solomon, S., Qin, D., Manning, M., Chen, Z., Marquis,

M., Averyt, K.B., Tignor, M., and H.L. Miller, Cambridge University Press, Cambridge, United Kingdom and New York, NY, USA, 996 pp.

Johnes, Penny J. 1996. Evaluation and management of the impact of land use change on the nitrogen and phosphorus load delivered to surface waters: the export coefficient modelling approach. *Journal of Hydrology* 183:323–349.

Kattaa, B., Al-Fares W., and A.R.A. Al Charideh. 2010. Groundwater vulnerability assessment for the Banyas Catchment of the Syrian coastal area using GIS and the RISKE method. *Journal of Environmental Management* 91:1103–1110.

McKinnney, D.C., Cai X., 2002. Linking GIS and water resources management models: an object-oriented method. *Environmental Modelling & Software* 17:413–425.

Mitof, I., Prăvălie, R., 2014. Temporal trends of hydroclimatic variability in the lower Buzău catchment. *Geographia Technica* 9(1):87–100.

Monteith, John L. 1965. "Evaporation and environment," In *Symposium of the Society for Experimental Biology. The State and Movement of Water in Living Organisms* edited by G.E. Fogg, 205–234, Vol. 19, Academic Press, Inc., NY.

Monteith, John L. 1985. "Evaporation from land surfaces: Progress in analysis and prediction since 1948" In *Advances in Evapotranspiration, Proc. National Conference on Advances in Evapotranspiration*, Dec. 16-17, Chicago, IL, Am. Soc. Agric. Eng., St. Joseph, MI, 4–12.

Nistor, Mărgărit M. 2014. *Using GIS techniques for the projection of ski tracks in Western Carpathians. Case study: Beliş district*. Cluj University Press, ISBN 978-973-595-709-4, Cluj-Napoca. http://www.editura.ubbcluj.ro/www/ro/ebooks/authors_d.php?ida=1068.

Nistor, M.M., Porumb-Ghiurco, C.G. 2015. How to compute the land cover evapotranspiration at regional scale? A spatial approach of Emilia-Romagna region. *GEOREVIEW Scientific Annals of Ştefan cel Mare University of Suceava Geography Series* 25(1):38–54.

Parmesan, C., Yohe, G. 2003. A globally coherent fingerprint of climate change impacts across natural systems. *Nature* 421(2):37–42.

Penman, Howard L. 1948. Natural evaporation from open water, bare soil and grass. *Proc. Royal Soc. London* 193(1032):120–145.

Rana, G., Katerji, N. 1998. A Measurement Based Sensitivity Analysis of the Penman-Monteith Actual Evapotranspiration Model for Crops of Different Height and in Contrasting Water Status. *Theor. Appl. Climatol.* 60:141–149.

Shaver, G.R., Canadell, J., Chapin, III F.S., Gurevitch, J., Harte, J., Henry, G., Ineson, P., Jonasson, S., Melillo, J., Pitelka, L., and L. Rustad. 2000. Global Warming and Terrestrial Ecosystems: A Conceptual Framework for Analysis. *Oxford Journals BioScience* 50(10):871–882.

Sumner, D.M., Jacobs, J.M. 2005. Utility of Penman–Monteith, Priestley–Taylor, reference evapotranspiration, and pan evaporation methods to estimate pasture evapotranspiration. *Journal of Hydrology* 308:81–104.

Stocks, B.J., Fosberg, M.A., Lynham, T.J., Mearns, L., Wotton, B.M., Yang, Q., Jin, J-Z., Lawrence, K., Hartley, G.R., Mason, J.A. and D.W. McKenney. 1998. Climate change and forest fire potential in Russian and Canadian boreal forests. *Climatic Change* 38:1–13.

The Canadian Centre for Climate Modelling and Analysis. 2014. The First Generation Coupled Global Climate Model Publishing Web, URL: http://www.ec.gc.ca/ccmac-cccma/default.asp?lang=En&n=540909E4-1. Accessed April 29, 2014.

Thornthwaite, Charles W. 1948. An approach toward a rational classification of climate. *Geographical Review* 38(1):55–94.

Valiantzas, John D. 2013. Simplified forms for the standardized FAO-56 Penman–Monteith reference evapotranspiration using limited weather data. *Journal of Hydrology* 505:13–23.

Yang, Xiaojun. 2005. Remote sensing and GIS applications for estuarine ecosystem analysis: an overview. *International Journal of Remote Sensing* 26(23):5347–5356.

BIBLIOGRAPHY

Back roads. Northern and Central Italy. LCCN: 2015207139 Main title: Back roads. Northern and Central Italy. Published/ Produced: London; New York: Dorling Kindersley, [2012]- Publication history: Began with 2012. Description: volumes: color illustrations, color maps; 23 cm + color maps (folded to 20 x 14 cm). Current frequency: Triennial LC classification: DG416 .B55 Variant title: Northern & Central Italy Related names: Dorling Kindersley, Inc, publisher. Subjects: Automobile travel--Italy, Northern--Guidebooks. Automobile travel--Italy, Central--Guidebooks. Automobile travel. Italy, Northern--Guidebooks. Italy, Central--Guidebooks. Italy--Italy, Central. Italy, Northern. Form/Genre: Guidebooks. Notes: Includes bibliographical references and index. Series: Eyewitness travel back roads

Documentary culture and the laity in the early Middle Ages LCCN: 2012017771 Main title: Documentary culture and the laity in the early Middle Ages / edited by Warren C. Brown, Marios Costambeys, Matthew Innes and Adam J. Kosto. Published/Produced: Cambridge: Cambridge University Press, 2013. Description: xvi, 389 pages: illustrations; 24 cm Links: Contributor biographical information http://www.loc.gov/catdir/enhancements/fy1210/2012017771-b.html Publisher description http://www.loc.gov/catdir/enhancements/fy1210/2012017771-d.html Table of contents only http://www.loc.gov/catdir/enhancements/fy1210/2012017771-t.html ISBN: 9781107025295 LC classification: D113 .D627 2013 Related names: Brown, Warren, 1963- Costambeys, Marios. Innes, Matthew. Kosto, Adam J.

Summary: “Many more documents survive from the early Middle Ages than from the Roman Empire. Although ecclesiastical archives may account for the dramatic increase in the number of surviving documents, this new investigation reveals the scale and spread of documentary culture beyond the Church. The contributors explore the nature of the surviving documentation without preconceptions to show that we cannot infer changing documentary practices from patterns of survival. Throughout Late Antiquity and the early Middle Ages - from North Africa, Egypt, Italy, Francia and Spain to Anglo-Saxon England - people at all social levels, whether laity or clergy, landowners or tenants, farmers or royal functionaries, needed, used and kept documents. The story of documentary culture in the early medieval world emerges not as one of its capture by the Church, but rather of a response adopted by those who needed documents, as they reacted to a changing legal, social and institutional landscape“-- Provided by publisher. Contents: Machine generated contents note: 1. Introduction; 2. Lay archives in the Late Antique and Byzantine East: the implications of the documentary papyri / Peter Sarris; 3. Public administration, private individuals and the written word in Late Antique North Africa, c.284-700 / Jonathan P. Conant; 4. Lay documents and archives in early Medieval Spain and Italy, c.400-700 / Nicholas Everett; 5. The gesta municipalia and the public validation of documents in Frankish Europe / Warren C. Brown; 6. Lay people and documents in the Frankish formula collections / Warren C. Brown; 7. Archives, documents and landowners in Carolingian Francia / Matthew Innes; 8. The production and preservation of documents in Francia: the evidence of cartularies / Hans Hummer; 9. The laity, the clergy, the scribes and their archives: the documentary record of eighth- and ninth-century Italy / Marios Costambeys; 10. Sicut mos esse solet: documentary practices in Christian Iberia, c.700-1000 / Adam J. Kosto; 11. On the material culture of legal documents: charters and their preservation in the Cluny archive, ninth to eleventh centuries / Matthew Innes; 12. Documentary practices, archives and lay people in central Italy (mid-ninth to eleventh centuries)

/ Antonio Sennis; 13. Archives and lay documentary practice in the Anglo-Saxon world / Charles Insley; 14. Conclusion. Subjects: Middle Ages--Sources. History--Sources. Civilization, Medieval--Sources. HISTORY / Europe / General. Europe--History--476-1492--Sources. Notes: Includes bibliographical references and index.

Dream destinations: the world's most unforgettable places LCCN: 2014397574 Personal name: Gallagher, Mary-Ann. Main title: Dream destinations: the world's most unforgettable places / Mary-Ann Gallagher. Published/Produced: New York: Metro Books, 2013. Description: 191 pages: color illustrations, color. maps; 25 cm ISBN: 9781435151659 LC classification: G153.4 .G35 2013 Contents: Giant's Causeway, Northern Ireland -- Stonehenge, England -- Alhambra, Granada, Spain -- Barcelona, Catalunya, Spain -- Mont-St-Michel, Normandy, France -- Loire Valley, Central France -- Paris, Northern France -- Bruges, West Flanders, Belgium -- Neuschwanstein, Bavaria, Germany -- Hammerfest, Finnmark, Norway -- Rome, Central Italy -- Venice, Veneto, Italy -- Metéora, Thessaly, Greece -- St Petersburg, Northwestern Russia -- Cappadocia, Central Anatolia, Turkey -- Petra, Arabah, Jordan -- Aswan to Luxor, Southern Egypt -- Skeleton Coast, Northern Namibia -- Victoria Falls, Zambia-Zimbabwe Border -- Samarkand, Central Uzbekistan -- Taj Mahal, Arga, India -- Darjeeling Railway, West Bengal, India -- Angkor Wat, Siem Reap, Cambodia -- Borobudur, Java, Indonesia -- Great Wall, Northern China -- Forbidden City, Beijing, China. Kyoto, Honshu, Japan -- Uluru, Northern Territory, Australia -- Great Barrier Reef, Queensland, Australia -- Tasmanian Wilderness, Australia -- Rapa Nui, Chile -- Vancouver, British Columbia, Canada -- Banff, Alberta, Canada -- Churchill, Manitoba, Canada -- Monument Valley, Utah, United States -- Yellowstone, Western United States -- Everglades, Florida, United States -- Washington Monuments, United States -- New England, Northeastern United States -- Mexican Day of the Dead -- Chichén Itzá, Yucatán Peninsula, Mexico -- Old Havana, Northwestern Cuba -- Cloud Forest, Costa Rica -- Galápagos, Ecuador -- Inca Trail, Near Cuzco, Peru -- Tierra del Fuego, Argentina and Chile -

- Perito Moreno Glacier, Argentina -- Manaus and the Amazon, Brazil -- Iguaçu Falls, Brazil-Argentina Border -- Haukadalur, Southern Iceland. Subjects: Travel--Guidebooks. Voyages and travels. Notes: Originally published: London: Quercus, 2010. Includes index.

Embodying value?: the transformation of objects in and from the ancient world LCCN: 2013487703 Main title: Embodying value?: the transformation of objects in and from the ancient world / edited by Annabel Bokern, Clare Rowan. Published/Produced: Oxford: Archaeopress, 2014. Description: v, 155 pages: illustrations (black and white), maps (black and white); 30 cm. ISBN: 9781407312200 (paperback) 1407312200 (paperback) LC classification: CC51 .E43 2014 Related names: Bokern, Annabel, editor of compilation. Rowan, Clare, 1982- editor of compilation. Momente de Transformation: Die Erzeugang und Zerstörung von Wert (Workshop) (2012) Embodying Value: The Transformation of Objects in and from the Roman World (Panel) (2012: Frankfurt am Main, Germany) Theoretical Roman Archaeology Conference (22nd: 2012: Universität Frankfurt am Main) Contents: Introduction: a Turkish teapot, or thinking things through / Annabel Bokern and Clare Rolwan -- Der Wert des Menschen und sein Preis: das Verhältnis von Person und Eigentum bei Brautpreis, Wergeld und Sklaverei / Emanuel Seitz -- Vergessen und neu belebt: das neolithische Steinbeil von seiner Produktion bis in die Neuzeit / Jennifer M. Bagley -- Transformers energize! Aegean Bronze Age rhyta in moments of transformation / Laerke Recht -- Der Wert des Kupfers: über die Entstehung und den Wandel von Wert vor 4000 Jahren / Martin Hensler -- Heavy metal in hallowed contexts: continuity and change in aes deposits in Central Italy and Sicily / Andreas M. Murgan -- The value of coinage in the Second Punic War and after / Clare Rowan -- Votives and values: communicating with the supernatural / Katherine M. Erdman -- The real value of fake: the curious case of cubiculum 34 in the house of Sallust (Pompeii VI 2, 4) / Suzanne van de Liefvoort -- The de(con)struction of public monuments / Stefanie Bauer -- Coin finds beyond the Danube:

functions of fourth century gold coins within barbarian societies / Dragana Eremić -- From manufactured goods to significant possessions: theorising pottery consumption in late antique Anatolia / William Anderson -- The power of things? Reconsidering the value of reused Roman material culture in the Medieval period of Serbia / Gordana Ciric. Subjects: Archaeology--Congresses. Romans--Europe--Congresses. Europe--Antiquities, Roman--Congresses. Rome--Antiquities--Congresses. Subject keywords: Ancient world Notes: Includes papers from a workshop entitled Momente de Transformation: Die Erzeugang und Zerstörung von Wert and a panel entitled Embodying Value: The Transformation of Objects in and from the Roman World held at the 22nd Theoretical Roman Archaeology Conference in Frankfurt. Both held in Spring 2012. Includes bibliographical references. Text in English and German; summaries in German and English. Series: BAR international series; 2592 BAR international series; 2592.

Fields, farms and colonists: intensive field survey and early Roman colonization in the Pontine region, central Italy LCCN: 2012455129 Personal name: Haas, Tymon C. A. de, 1979- Main title: Fields, farms and colonists: intensive field survey and early Roman colonization in the Pontine region, central Italy / Tymon C.A. de Haas. Published/Created: Eelde: Barkhuis; Groningen: Groningen University Library, 2011. Description: 2 v. (xi, 487 p.): ill. (some col.), maps; 30 cm. ISBN: 9789077922934 (pbk.) 9077922938 (pbk.) LC classification: DG55.P66 H33 2011 Related names: Rijksuniversiteit te Groningen. Bibliotheek. Contents: v. 1. Text -- v. 2. Catalogue. Subjects: Excavations (Archaeology)--Italy--Pontine Marshes. Kolonisten. Romeinen (volk) Veldwerk. Pontine Marshes (Italy)--Antiquities, Roman. Lazio. Form/Genre: Proefschriften (vorm) Notes: Originally presented as the author's thesis (Ph.D)--Rijksuniversiteit Groningen. Published in association with Groningen University Library. Includes bibliographical references. In English with a summary in Dutch. Additional formats: Fields, farms and colonists (NL-LeOCL)339152230 Series: Groningen archaeological studies, 1572-1760; v. 15

Groningen archaeological studies; v. 15.

Fingerprinting the Iron Age: approaches to identity in the European Iron Age: integrating South-Eastern Europe into the debate LCCN: 2014034508 Main title: Fingerprinting the Iron Age: approaches to identity in the European Iron Age: integrating South-Eastern Europe into the debate / Catalin Nicolae Popa and Simon Stoddart. Published/Produced: Oxford; Philadelphia: Oxbow Books, 2014. Description: xii, 428 pages; 25 cm ISBN: 9781782976752 LC classification: GN780.2.A1 F56 2014 Related names: Popa, Cătălin Nicolae, editor, author. Stoddart, Simon, editor, author. Contents: Introduction: the challenge of Iron Age identity / Simon Stoddart and Catalin Nicolae Popa -- Part I. Perspectives from south east Europe -- The coexistence and interference of the late iron age Transylvanian communities / Sándor Berecki -- Identities of the early Iron Age in north-eastern Slovenia / Matija Crenar and DimitrijMleku -- Royal bodies, invisible victims: gender in the funerary record of late Iron Age and early Hellenistic Thrace / Bela Dimova -- Mediterranean wine and Dacian conviviality: ancient and modern myths and archaeological evidence / Mariana Egri -- Sarmizegetusa regia: the identity of a royal site? / Gelu Florea -- The ethnic construction of early Iron Age burials in Transylvania: Scythians, Agathyrsi or Thracians? / Alexandra Ghenghea -- Negotiating identities at the edge of the Roman Empire / Marko A. Jankovic -- Tracing ethnicity backwards: the case of the central Balkan tribes / Vladimir D. Mihajlovic -- The quest for group identity in late Iron Age Romania: statistical reconstruction of groups based on funerary evidence / Catalin Nicolae Popa -- Changing identities of the Iron Age communities of southern Pannonia / Yvonne Inall, Hrvoje Potrebica and Marko Dizdar -- Indigenous and colonist communities in the eastern Carpathianbasin at the beginning of the late Iron Age: the genesis of an eastern Celtic world / Aurel Rustoiu -- Ancient Thrace between the east and the west / Nikola Theodossiev -- "Hellenisation" and ethnicity in the continental Balkan Iron Age / Ivan Vranic -- Part II. Perspectives from the west -- Central places and the construction of collective identities in the middle Rhine-Moselle region / Manuel Fernández-Götz -- Fingerprinting Iron Age communities in south-west Germany and an integrative theory of culture / Oliver Nakoinz -- Iron Age identities in central Europe: some initial approaches / Peter C. Ramsl -- Part III. Perspectives from the far west -- Negotiating identity on the edge of empire / Louisa Campbell -- Personal adornment in Iron Age Britain: the case of the missing glass beads / Elizabeth Foulds -- Part IV. Perspectives from the south west -- Spoiling for a fight: using spear typologies to identify aspects of warrior identity and fighting style in Iron Age south Italy -- Communal vs. individual: the role of identity in the burials of Peucetia / Olivia Kelley -- A view from the south (west): identity in Tyrrhenian central Italy / Simon Stoddart -- Part V. Synthesis -- Identity, integration, power relations and the study of the European Iron Age: implications from Serbia / Staa Babic -- The Celts: more myths and inventions / John Collis -- Material culture and identity: the problem of identifying Celts, Germans and Romans in late Iron Age Europe / Peter Wells -- Fingerprinting the European Iron Age: historical, cultural and intellectual perspectives on identity and ethnicity / Catalin Nicolae Popa and Simon Stoddart. Subjects: Iron age--Europe. Social archaeology--Europe. Group identity--Europe--History. Social structure--Europe--History. Europe--Antiquities. Notes: Includes bibliographical references and index. Additional formats: Online version: Fingerprinting the Iron Age Oxford: Oxbow Books, 2014 9781782976769 (DLC) 2014044700

For the love of Italy: rural pleasures and hotel estates LCCN: 2009047508 Personal name: Caracciolo Chia, Marella. Main title: For the love of Italy: rural pleasures and hotel estates / Marella Caracciolo; photographs by Oberto Gili. Edition: 1st ed. Published/Created: New York: Clarkson Potter/Publishers, c2010. Description: 224 p.: col. ill.; 32 cm. Links: Contributor biographical information http://www.loc.gov/catdir/enhancements/fy1112/2009047508-b.html Publisher description http://www.loc.gov/catdir/enhancements/fy1112/2009047508-d.html ISBN: 9780307452481 0307452484 LC classification:

DG430.2 .C35 2010 Related names: Gili, Oberto. Contents: Introduction -- Northern Italy -- Tuscany -- Central Italy -- Insider's list. Subjects: Agritourism--Italy. Farms--Recreational use--Italy. Italy--Description and travel. Italy--Pictorial works. Notes: Includes index.

History of Italian Renaissance art: painting, sculpture, architecture LCCN: 2009033197 Personal name: Hartt, Frederick. Main title: History of Italian Renaissance art: painting, sculpture, architecture / Frederick Hartt, David G. Wilkins. Edition: 7th ed. Published/Created: Upper Saddle River: Prentice Hall, c2011. Description: 736 p.: ill. (chiefly col.), col. maps; 30 cm. ISBN: 9780205705818 (pbk.) 0205705812 (pbk.) 9780205707973 (hardback) 0205707971 (hardback) LC classification: N6915 .H37 2011 Related names: Wilkins, David G. Contents: Italy and Italian art -- The late Middle Ages -- Duecento art in Tuscany and Rome -- Florentine art of the early trecento -- Sienese art of the early trecento -- Later Gothic art in Tuscany and northern Italy -- The quattrocento -- The Renaissance begins: architecture -- Transitions in Tuscan sculpture -- Transitions in Florentine painting -- The heritage of Masaccio: Fra Angelico and Fra Filippo Lippi -- Florentine architecture and sculpture: c. 1430-1455 -- Florentine painting at mid-century -- Art in Florence under the Medici I -- Art in Florence under the Medici II -- The Renaissance in central Italy -- Gothic and Renaissance in Venice and northern Italy -- The cinquecento -- The origins of the high Renaissance --The high Renaissance in Rome -- New developments, c. 1520-50 -- High and late Renaissance in Venice and on the mainland -- The late sixteenth century Subjects: Art, Italian. Art, Renaissance--Italy. Notes: Includes bibliographical references (p. 700-714) and index.

In search of the Samnites: adornment and identity in archaic central Italy, 750-350 B.C. LCCN: 2013432683 Personal name: Richardson, Amy. Main title: In search of the Samnites: adornment and identity in archaic central Italy, 750-350 B.C. / Amy Richardson. Published/Created: Oxford: Archaeopress, 2013. Description: xi, 162 p.: ill., maps; 30 cm. ISBN: 9781407311708 1407311700 LC classification: DG225.S26 R53 2013 Subjects: Samnites--Material culture. Sannio (Italy)--Antiquities. Notes: Originally presented as: Thesis (Ph.D.)--University of Reading, 2010. Includes bibliographical references. Series: BAR international series; 2550 BAR international series; 2550.

Magnetic methods and the timing of geological processes LCCN: 2013370283 Main title: Magnetic methods and the timing of geological processes / edited by L. Jovane, Universidade de São Paulo, Brazil, E. Herrero-Bervera, University of Hawaii at Manoa, USA, L. A. Hinnov, Johns Hopkins University, USA, and B. Housen, Western Washington University, USA. Published/Produced: London: Geological Society, 2013. Description: vi, 402 pages: illustrations (some color), maps (some color); 26 cm ISBN: 9781862393547 1862393540 LC classification: QE508 .M29 2013 Related names: Jovane, Luigi. Herrero-Bervera, Emilio. Hinnov, L. A. (Linda Alide) Housen, Bernard Arthur. Abstract: Magnetostratigraphy is best known as a technique that employs correlation among different stratigraphic sections using the magnetic directions defining geomagnetic polarity reversals as marker horizons. The ages of the polarity reversals provide common tie points among the sections, allowing accurate time correlation. Recently, studies of magnetic methods and the timing of geological processes have acquired a broader meaning, now referring to many types of magnetic measurements within a stratigraphic sequence. Many of these measurements provide correlation and age control not only for the older and younger boundaries of a polarity interval, but also within intervals. Thus, magnetostratigraphy no longer represents a dating tool based only on geomagnetic polarity reversals, but comprises a set of techniques that includes measurements of geomagnetic field parameters, environmental magnetism, rock-magnetic properties, radiometric dating and astronomically forced palaeoclimatic change recorded in sedimentary rocks, and key corrections to magnetic directions related to geodynamics, palaeocurrents, tectonics and diagenetic processes -- Source other than

Library of Congress. Contents: Magnetic methods and the timing of geological processes / L. Jovane, L. Hinnov, B.A. Housen and E. Herrero-Barvera -- Oligocene-Miocene magnetostratigraphy of deep-sea sediments from the equatorial Pacific (IODP Site U1333) / E.P. Guidry, C. Richter, G.D. Acton, J.E.T. Channell, H.F. Evans, C. Ohneiser, Y. Yamamoto and T. Yamazaki -- Integrated biomagnetochronology for the Palaeogene of ODP Hole 647A: implications for correlating palaeoceanographic events from high to low latitudes / J.V. Firth, J.S. Eldrett, I.C. Harding, H.K. Coxall and B.S. Wade -- Integrated magnetobiostratigraphy of the middle Eocene-lower Oligocene interval from the Monte Cagnero section, central Italy / L. Jovane, J.F. Savian, R. Coccioni, F. Frontalini, G. Bancalà, R. Catanzariti, V. Luciani, S.M. Bohaty, P.A. Wilson and F. Florindo -- Middle Eocene to early Oligocene magnetostratigraphy of ODP Hole 711A (Leg 115), western equatorial Indian Ocean / J.F. Savian, L. Jovane, S.M. Bohaty, and P.A. Wilson -- Integrated stratigraphy (magneto-, bio- and chronostratigraphy) and geochronology of the Palaeogene pelagic succession of the Umbria-Marche Basin (central Italy) / R. Coccioni, M. Sideri, G. Bancalà, R. Catanzariti, F. Frontalini, L. Jovane, A. Montanari, and J. Savian -- Magnetostratigraphy, fence diagrams and basis analysis / M. Yan, R. Van Der Voo, X.-M. Fang and C. Song -- Oligocene slow and Miocene-Quaternary rapid information and uplift of the Yumu Shan and North Qilian Shan: evidence from high-resolution magnetostratigraphy and tectonosedimentology / X. Fang, D. Liu, C. Song, S. Dai and Q. Meng -- Neogene rotations in the Jiuquan Basin, Hexi Corridor, China / M. Yan, X. Fang, R. Van Der Voo, C. Song and J. Li -- Magnetostratigraphic results from sedimentary rocks of IODP's Nankai Trough Seismogenic Zone Experiment (NanTroSEIZE) Expedition 322 / X. Zhao, H. Oda, H. Wu, T. Yamamoto, Y. Yamamoto, Y. Yamamoto, T. Nakajima, Y. Kitmura and T. Kanamatsu -- Evaluating Late Holocene radiocarbon-based chronologies by matching palaeomagnetic secular variations to geomagnetic field models: an example from Lake Kalimpaa (Sulawesi, Indonesia) / T. Haberzettl, G. St-Onge, H.

Behling and W. Kirleis -- On the directional geomagnetic signature of the Pringle Falls excursion recorded at Pringle Falls, Oregon, USA / E. Herrero-Bervera and E. Cañón-Tapia -- On the palaeomagnetic and rock magnetic constraints regarding the age of IODP 325 Hole M0058A / E. Herrero-Bervera and L. Jovane -- Characteristic wavelengths in VGP trajectories from magnetostratigraphic data of the Early Cretaceous Serra Geral lava piles, southern Brazil / G. Caminha-Maciel and M. Ernesto -- Rock-magnetic cyclostratigraphy for the Late Pliocene-Early Pleistocene Stirone section, Northern Apennine mountain front, Italy / K.L. Gunderson, K.P. Kodama, D.J. Anastasio and F.J. Pazzaglia -- Global Milankovitch cycles recorded in rock magnetism if the shallow marine lower Cretaceous Cupido Formation, northeastern Mexico / L.A. Hinnov, K.P. Kodama, D.J. Anastasio, M.Elrick and D.K. Latta -- Visual identify-cation and quantification of Milankovitch climate cycles in outcrop: an example from the Upper Ordovician Kope Formation, Northern Kentucky / B.B. Ellwood, C.E. Brett, J.H. Tomkin and W.D. MacDonald -- Anisotropy of magnetic susceptibility and sedimentary cycle data from Permo-Carboniferous rhythmites (Paraná Basin, Brazil): a multiple proxy record of astronomical and millennial scale palaeoclimate change in a glacial setting / D.R. Franco and L.A. Hinnov -- Magnetostratigraphy susceptibility for the Guadalupian series GSSPs (Middle Permian) in Guadalupe Mountains National Park and adjacent areas in West Texas / B.B. Ellwood, L.L. Lambert, J.H. Tomkin, G.L. Bell, M.K. Nestell, G.P. Nestell and B.R. Wardlaw. Subjects: Geochronometry. Notes: Includes bibliographical references and index. Series: Geological Society special publication, 0305-8719; no. 373 Geological Society special publication; no. 373.

Monumentality in Etruscan and early Roman architecture: ideology and innovation LCCN: 2011048877 Main title: Monumentality in Etruscan and early Roman architecture: ideology and innovation / edited by Michael L. Thomas and Gretchen E. Meyers. Published/Produced: Austin: University of Texas Press,

[2012] ©2012 Description: 184 pages: illustrations; 27 cm ISBN: 9780292738881 (cloth: alk. paper) 9780292738898 (e-book) LC classification: NA300 .M66 2012 Related names: Thomas, Michael L., 1966- editor of compilation. Meyers, Gretchen E., 1970- author, editor of compilation. Edlund-Berry, Ingrid E. M., author, honouree. Contents: Introduction: the experience of monumentality in Etruscan and early Roman architecture / Gretchen E. Meyers -- Straw to stone, huts to houses: transitions in building practices and society in protohistoric Latium / Elizabeth Colantoni -- The performance of death: monumentality, burial practice, and community identity in central Italy's urbanizing period / Anthony Tuck -- Monumentalization of the Etruscan round moulding in sixth-century BCE central Italy / Nancy A. Winter -- Monumental embodiment: somatic symbolism and the Tuscan temple / P. Gregory Warden -- The Capitoline temple and the effects of monumentality on Roman temple design / John N. Hopkins -- On the introduction of stone entablatures in republican temples in Rome / Penelope J. E. Davies -- Afterword. Reflections / Ingrid E. M. Edlund-Berry. Subjects: Architecture, Etruscan. Architecture, Roman--Italy, Central. Notes: Includes index.

Processes of cultural change and integration in the Roman world LCCN: 2015011626 Main title: Processes of cultural change and integration in the Roman world / edited by Saskia T. Roselaar. Published/Produced: Leiden; Boston: Brill, [2015] Description: x, 314 pages: illustrations, maps; 25 cm. ISBN: 9789004294547 (hardback: alkaline paper) LC classification: DG87 .P76 2015 Related names: Roselaar, Saskia T., author editor of compilation. Summary: “Processes of Cultural Change and Integration in the Roman World is a collection of studies on the interaction between Rome and the peoples that became part of its Empire between c. 300 BC and AD 300. The book focuses on the mechanisms by which interaction between Rome and its subjects occurred, e.g., the settlements of colonies by the Romans, army service, economic and cultural interaction. In many cases Rome exploited the economic resources of the conquered territories without allowing the local inhabitants any legal autonomy. However, they usually maintained a great deal of cultural freedom of expression. Those local inhabitants who chose to engage with Rome, its economy and culture, could rise to great heights in the administration of the Empire. Contributors are: Patricia Argüelles, Aitor Blanco-Peréz, Elisabeth Buchet, Christopher Burden-Strevens, Tamara Dijkstra, Leonardo Gregoratti, Maurizio Gualtieri, Alfred Hirt, Enora Le Quéré, Josipa Lulić, Daniele Miano, Alexander Rubel, Rafael Scopacasa, Christopher Sparey-Green, Marleen Termeer, and Fiona Tweedie"-- Provided by publisher. Contents: Introduction: processes of cultural change and integration in the Roman world / Saskia T. Roselaar -- Theorizing Romanization: cognition and cultural change in Roman provinces: a case of religious change in Roman Dalmatia / Josipa Lulić -- An allied view of integration: Italian elites and consumption in the second century BC / Rafael Scopacasa -- Minting apart together: bronze coinage production in Campania and beyond in the third century BC / Marleen Termeer -- The archaeology of 'integration' in western Lucania: a review of recent work / Maurizio Gualtieri -- Volaterrae and the gens Caecina / Fiona C. Tweedie -- Iniungi delectus: the recruitment of Britons in the Roman army during the conquest: the evidence from Dorset / Christopher Sparey-Green -- Apamea and the integration of a Roman colony in western Asia Minor / Aitor Blanco-Pérez -- Burial and commemoration in the Roman colony of Patras / Tamara Dijkstra -- Akkulturation und Integration in der römischen Dobruscha: das Fallbeispiel der römischen Siedlung Ibida (Slava Rusă) in Rumänien / Alexander Rubel -- Roman exploitation and new road infrastructures in Asturia Transmontana (Asturias, Spain) / Patricia A. Argüelles Álvarez -- Mines and economic integration of provincial 'frontiers' in the Roman Principate / Alfred M. Hirt -- The 'opportunistic exploitation' of Melos: a case study of economic integration and cultural change in the Roman Cyclades / Enora Le Quéré -- Roman traders as a factor of Romanization in Noricum and in the eastern transalpine region / Leonardo Gregoratti -- Spreading virtues in republican Italy / Daniele Miano -- Literary topoi and the integration of central Italy / Elisabeth Buchet -- 'Ein völlig romanisierter Mann'?: identity, identification, and integration in the Roman History of Cassius Dio and in Arrian / Christopher Burden-Strevens. Subjects: Roman provinces--Administration--Congresses. Roman provinces--Social conditions--Congresses. Indigenous peoples--Rome--Provinces--History--Congresses. Assimilation (Sociology)--Rome--Congresses. Acculturation--Rome--Congresses. Rome--History--Republic, 265-30 B.C.--Congresses. Rome--History--Empire, 30 B.C.-476 A.D.--Congresses. Notes: This volume is the result of a conference held at the University of Nottingham in July 2013, which focused on processes of integration in the Roman world. This meeting was a follow-up to an earlier conference, held at Manchester in 2010, which looked at processes of integration in the Roman Republic (see LCCN 2012007861). Both conferences started from the idea that, despite the amount of recent scholarship on integration in the ancient world and the impact these had on formation of identities, there are still aspects of these issues that are not fully understood. Includes bibliographical references and index. Collection of 17 essays, with 16 in English and 1 in German. Series: Mnemosyne supplements. History and archaeology of classical antiquity, 2352-8656; volume 382

Recent prehistoric enclosures and funerary practices in Europe: proceedings of the international meeting held at the Gulbenkian Foundation (Lisbon, Portugal, November 2012) LCCN: 2014486327 Meeting name: Recent Prehistoric Enclosures and Funerary Practices in Europe (Meeting) (2012: Lisbon, Portugal) Main title: Recent prehistoric enclosures and funerary practices in Europe: proceedings of the international meeting held at the Gulbenkian Foundation (Lisbon, Portugal, November 2012) / edited by António Carlos de Valera. Published/Produced: Oxford: Archaeopress, 2014. Description: iii, 154 pages: illustrations (black and white), maps (black and white); 30 cm. ISBN: 9781407313184 (paperback) 1407313185 (paperback) LC classification: GT3242 .R43 2012 Related names: Valera, António Carlos, editor. Contents: The times and timings of enclosures / Alasdair

Whittle -- Enclosures & burial in middle & late Neolithic Britain / Alex Gibson -- The place of human remains and funerary practices in recent neolithic ditched and walled enclosures in the west of France (IV-III Mill. BC) / Audrey Blanchard, Jean-Noël Guyodo, Ludovic Soler -- Funerary practices and body manipulation at Neolithic and Chalcolithic Perdigões ditched enclosures (South Portugal) / António Carlos Valera, Ana Maria Silva, Claudia Cunha, Lucy Shaw Evangelista -- Skeletons in the ditch: funerary activity in ditched enclosures of Porto Torrão (Ferreira do Alentejo, Beja) / Filipa Rodrigues -- Enclosures and funerary practices: about an archaeology in search for the symbolic dimension of social relations / Susana Oliveira Jorge -- Human bones from Chalcolithic walled enclosures of Portuguese estremadura: the examples of Zambujal and Leceia / Michael Kunst, João Luís Cardoso, Anna Waterman -- Human sacrifices with cannibalistic practices in a pit enclosure? The extraordinary early Neolithic site of Herxheim (Palatinate, Germany) / Andrea Zeeb-Lanz -- Gendered burials at an henge-like enclosure near Magdeburg, central Germany: a tale of reverence and ritual killing? / André Spatzier Marcus Stecher, Kurt W. Alt. François Bertemes -- The Copper age ditched settlement at Conelle de Arcevia (Central Italy) / Alberto Cazzella, Giulia Recchia -- Funerary practices in the ditched enclosures of Camino de las Yeseras: ritual, temporal and spatial diversity / Patrícia Rios, Corina Liesau, Concepción Blasco -- Recent prehistory enclosures & funerary practices / José Enrique Márquez Romero, Vítor Jímenez Jaímez. Subjects: Funeral rites and ceremonies--Europe--Congresses. Excavations (Archaeology)--Europe--Congresses. Burial--Europe--Congresses. Prehistoric peoples. Europe. Subject keywords: Enclosures Funerary practices Notes: Includes bibliographical references. Series: BAR international series; 2676 BAR international series; 2676.

Religious poverty, visual riches: art in the Dominican churches of Central Italy in the thirteenth and fourteenth Centuries LCCN: 2013005142 Personal name: Cannon, Joanna. Main title: Religious poverty, visual riches: art in the Dominican churches of Central Italy in the thirteenth and fourteenth Centuries /

Joanna Cannon. Published/Produced: New Haven: Yale University Press, [2013] Description: xi, 444 pages: illustrations (some color); 29 cm ISBN: 9780300187656 (cl: alk. paper) LC classification: NK2192.I82 C463 2013 Subjects: Dominicans--Art collections. Church decoration and ornament--Italy, Central--History--To 1500. Dominican architecture--Italy, Central. Christian art and symbolism--Italy, Central--Medieval, 500-1500. Art and society--Italy, Central--History--To 1500. Architecture and society--Italy, Central--History--To 1500. Notes: Includes bibliographical references and index.

Republics at war, 1776-1840: revolutions, conflicts, and geopolitics in Europe and the Atlantic world LCCN: 2013042166 Main title: Republics at war, 1776-1840: revolutions, conflicts, and geopolitics in Europe and the Atlantic world / edited by Pierre Serna, Antonino De Francesco, and Judith Miller. Published/Produced: Houndmills, Basingstoke, Hampshire: Palgrave Macmillan, 2013. Description: xvi, 291 pages: maps; 23 cm. ISBN: 9781137328816 (hardback) LC classification: D309 .R46 2013 Related names: Serna, Pierre. De Francesco, Antonino, 1954- Miller, Judith A., 1956- Scope and content: "This book places itself at the intersection of two fields of study--military history and political ideologies--in order to investigate the troubling links between warfare and republicanism during the Revolutionary era. This international team of historians probes the dynamics of nations born of revolutions, and the violent confrontations that erupted as republicans carried their principles beyond their borders. The collection presents fresh work, including articles by scholars who have not previously published in English. Their wide-ranging inquiries highlight the impact of war on slave emancipation in the Caribbean and the United States, as well as the attempts to impose republicanism through warfare in Ireland, Italy and Spain. They trace debates in theaters, diplomatic communiques, and conscription strategies to understand the meaning of war in the name of a republic. Together, the contributions reveal the profound, often damaging, and sometimes liberating consequences of those combined military and political undertakings"-- Provided by publisher. Contents: Introduction: War and the Republic: Dangerous Liaisons / Pierre Serna -- PART I. CALLING FOR "REPUBLICAN" WAR -- The American Origins of the French Revolutionary War / Antonino De Francesco -- In Search of a Glorious Peace? Republican Diplomats at War, 1792-1799 / Virginie Martin -- Can a Powerful Republic Be Peaceful? The Debate of Year IV on the Place of France in the European Order / Marc Belissa -- Theobald Wolfe Tone's Mission to France 1796-1798 / Sylvie Kleinman -- PART II. CITIZENSHIP AND "REPUBLICAN" WAR -- A "Black Declaration of Independence"? War, Republic, and Race in the United States of America, 1775-1787 / Marie-Jeanne Rossignol -- The Army of the Republic: New warfare and a new army / Annie Crépin -- A Patriotic School: The Recruitment of the Italian Legion in France, 1799-1800 / Katia Visconti -- From Individual to Collective Emancipation: War and the Republic in the Caribbean during the French Revolution / Frédéric Régent -- PART III. REJECTING "REPUBLICAN" WAR -- Fratricide: Tragic Brothers, Masculine Violence and the Republic on the French Stage, 1799 / Judith A. Miller -- War and Citizenship: Central Italy, 1798-1799 / Bernard Gainot -- The Strength of Weapons, the Strength of Opinions: Counter-Revolution in the Papal States, 1790-1799 / Mario Tosti -- International War, National War, Civil War: Spain and Counterrevolution, 1793-1840 / Pedro Rújula. Subjects: Republicanism--History--18th century. Republicanism--History--19th century. Revolutions--History--18th century. Revolutions--History--19th century. Geopolitics--History--18th century. Geopolitics--History--19th century. HISTORY / Europe / General. HISTORY / Military / Other. HISTORY / Modern / 18th Century. HISTORY / Modern / 19th Century. HISTORY / Social History. Europe--History, Military--18th century. Europe--History, Military--19th century. United States--History, Military--18th century. United States--History, Military--19th century. Notes: Includes bibliographical references.

Rubens: copies and adaptations from Renaissance and later artists.

Italian artists. 3, Artists working in central Italy and France LCCN: 2012454861 Personal name: Wood, Jeremy, 1950- Main title: Rubens: copies and adaptations from Renaissance and later artists. Italian artists. 3, Artists working in central Italy and France / by Jeremy Wood. Published/Created: London: Harvey Miller, c2011. Description: 2 v.: ill.; 27 cm. ISBN: 9781905375417 (hbk.: set) 1905375417 (hbk.: set) 9781905375790 (v. 1) 1905375794 (v. 1) 9781905375806 (v. 2) 1905375808 (v. 2) LC classification: N6973.R9 W663 2011 Portion of title: Copies and adaptations from Renaissance and later artists: Italian artists Artists working in central Italy and France Italian artists Contents: v. 1. [text] -- v. 2. [illustrations]. Subjects: Rubens, Peter Paul, 1577-1640--Catalogues raisonnés. Rubens, Peter Paul, 1577-1640--Sources. Art, Italian--Italy, Central--Copying. Art, French--Copying. Art--Adaptations. Notes: Includes bibliographical references and indexes. Series: Corpus Rubenianum Ludwig Burchard; pt. 26 (2), III Corpus Rubenianum Ludwig Burchard; pt. 26 (2), III.

Seafloor geomorphology as benthic habitat: GeoHAB atlas of seafloor geomorphic features and benthic habitats LCCN: 2011943671 Main title: Seafloor geomorphology as benthic habitat: GeoHAB atlas of seafloor geomorphic features and benthic habitats / edited by Peter T. Harris, Elaine K. Baker. Edition: 1st ed. Published/Created: Amsterdam; Boston: Elsevier, 2012 Description: xlv, 900 p.: ill., maps; 24 cm. ISBN: 9780123851406 0123851408 LC classification: GC87 .S43 2012 Portion of title: GeoHAB atlas of seafloor geomorphic features and benthic habitats Related names: Harris, Peter T. (Peter Townsend), 1958- Baker, Elaine K., 1957- GeoHab 2007 (2007: Nouméa, New Caledonia) Contents: Why map benthic habitats? / Peter T. Harris and Elaine K. Baker -- Habitat mapping and marine management / Elaine K. Baker and Peter T. Harris -- Anthropogenic threats to benthic habitats / Peter T. Harris -- Biogeography, benthic ecology, and habitat classification schemes / Peter T. Harris -- Surrogacy / Peter T. Harris -- Seafloor geomorphology: coast, shelf, and abyss / Peter T. Harris -- Phanerogam meadows: a characteristic habitat of the Mediterranean shelf: examples from the Tyrrhenian Sea / Silvana D'Angelo and Andrea Fiorentino -- Predictive modeling of dominant macroalgae abundance on temperate island shelves (Azores, northeast Atlantic) / Fernando Tempera ... [et al.] -- Methane-related carbonate cementation of marine sediments and related macroalgal coralligenous assemblages in the northern Adriatic Sea / Emiliano Gordini ... [et al.] -- Coastal kelp forest habitat in the Baie des Chaleurs, Gulf of St. Lawrence, Canada / Antoine Collin, Bernard Long, and Philippe Archambault -- The Wadden Sea in the Netherlands: ecotopes in a World Heritage barrier island system / Norbert Dankers ... [et al.] -- Sand wave field: the OBel Sands, Bristol Channel, UK / J.W. Ceri James ... [et al.] -- Benthic habitat variations over tidal ridges, North Sea, the Netherlands / Thaiënne A.G.P. van Dijk ... [et al.] -- Fine-scale geomorphological mapping of sandbank environments for the prediction of macrobenthic occurrences, Belgian part of the North Sea / Vera Van Lancker ... [et al.] -- Large submarine sand waves and gravel lag substrates on Georges Bank off Atlantic Canada / Brian J. Todd and Page C. Valentine -- The Yongala's "halo of holes": systematic bioturbation close to a shipwreck / Thomas C. Stieglitz -- Submarine De Geer moraines in the Kvarken Archipelago, the Baltic Sea / Aarno T. Kotilainen ... [et al.] -- Habitats and benthos of an evolving fjord, Glacier Bay, Alaska / Guy R. Cochrane ... [et al.] -- Geomorphic features and benthic habitats of a sub-Arctic fjord: Gilbert Bay, southern Labrador, Canada / Alison Copeland ... [et al.] -- Seabed character and habitats of a rocky Antarctic coastline: Vestfold Hills, East Antarctica / Philip E. O'Brien ... [et al.] -- Fringing reefs of the Seychelles Inner granitic islands, western Indian Ocean / Jenny S. Collier and Stuart R. Humber -- Coral reefs and reef islands of the Amirantes Archipelago, western Indian Ocean / Sarah Hamylton, Tom Spencer and Annelise B. Hagan -- Hyperspectral remote sensing of the geomorphic features and habitats of the AI Wajh Bank reef system, Saudi Arabia, Red Sea / Sarah Hamylton -- Mesophotic coral reefs of the Puerto Rico shelf / Roy A. Armstrong and Hanumant Singh -- Geomorphic features and infauna diversity of a subtropical mid-ocean carbonate shelf: Lord Howe Island, southwest Pacific Ocean / Brendan P. Brooke ... [et al.] -- Geomorphology of reef fish spawning aggregations in Belize and the Cayman Islands (Caribbean) / William D. Heyman and Shinichi Kobara -- Submerged reefs and aeolian dunes as inherited habitats, Point Cloates, Carnarvon shelf, Western Australia / Scott L. Nichol ... [et al.] -- Seafloor morphology and coral habitat variability in a volcanic environment: Kaloko-Honokohau National Historical Park, Hawaii, USA / Ann E. Gibbs and Susan A. Cochran -- Habitats and benthos at Hydrographers Passage, Great Barrier Reef, Australia / Robin J. Beaman ... [et al.] -- Two shelf-edge marine protected areas in the eastern Gulf of Mexico / Rebecca J. Allee, Andrew W. David, and David F. Naar -- Nontropical carbonate shelf sedimentation: the Archipelago Pontino (central Italy) case history / Eleonora Martorelli ... [et al.] -- Habitats of the Cap de Creus continental shelf and Cap de Creus canyon, northwestern Mediterranean / Claudio Lo Iacono ... [et al.] -- Rock ridges in the central English Channel / Roger A. Coggan and Markus Diesing -- Characterization of shallow inshore coastal reefs on the Tasman Peninsula, southeastern Tasmania, Australia / Vanessa Lucieer ... [et al.] -- Rocky reef and sedimentary habitats within the continental shelf of the southeastern Bay of Biscay / Ibon Galparsoro ... [et al.] -- Rock reefs of British Columbia, Canada: inshore rockfish habitats / K. Lynne Yamanaka ... [et al.] -- Seabed habitats of the southern Irish Sea / Karen A. Robinson ... [et al.] -- Habitats and demersal fish communities in the vicinity of Albatross Bank, Gulf of Alaska / Jennifer R. Reynolds ... [et al.] -- Seabed habitat of a glaciated shelf, German Bank, Atlantic Canada / Brian J. Todd, Vladimir E. Kostylev, and Stephen J. Smith -- Identifying potential habitats from multibeam echosounder imagery to estimate abundance of groundfish: a case study at Heceta Bank, OR, USA / Julia E.R. Getsiv-Clemons ... [et al.] -- Open shelf valley system, northern palaeovalley, English Channel, UK / J.W. Ceri James ... [et al.] -- Benthos supported by the tunnel-valleys of the southern North Sea / Bryony Pearce ... [et al.] -- Benthic habitats and benthic communities in southeastern Baltic Sea, Russian sector /

Elena Ezhova ... [et al.] -- Inland tidal sea of the northeastern Pacific / J. Vaughn Barrie ... [et al.] -- Cold-water coral distribution in an erosional environment: the Strait of Gibraltar gateway / Ben De Mol ... [et al.] -- Habitat mapping of a cold-water coral mound on Pen Duick escarpment (Gulf of Cadiz) / Lies De Mol ... [et al.] -- Habitats at the Rockall Bank slope failure features, northeast Atlantic Ocean / Janine Guinan ... [et al.] -- Evaluating geomorphic features as surrogates for benthic biodiversity oil on Australia's western continental margin / Franziska Althaus ... [et al.] -- Habitats of the Chella Bank, eastern Alboran Sea (western Mediterranean) / Claudio Lo Iacono ... [et al.] -- Habitat heterogeneity in the Nazaré deep-sea canon offshore Portugal / Veerle A.I. Huvenne ... [et al.] -- Banks, troughs, and canyons on the continental margin off Lofoten, Vesterålen, and Troms, Norway / Lene Buhl-Mortensen ... [et al.] -- Distribution of hydrocorals along the George V slope, East Antarctica / Alexandra L. Post ... [et al.] -- The Cook Strait canyon, New Zealand: geomorphology and seafloor biodiversity of a large bedrock canyon system in a tectonically active environment / Geoffroy Lamarche ... [et al.] -- The Ascension-Monterey canyon system: habitats of demersal fishes and macroinvertebrates along the central California coast of the USA / Mary Yoklavich and H. Gary Greene -- A study of geomorphological features of the seabed and the relationship to deep-sea communities on the western slope of Hatton Bank (NE Atlantic Ocean) / Miriam Sayago-Gil ... [et al.] -- Seafloor habitats and benthos of a continental ridge: Chatham rise, New Zealand / Scott D. Nodder ... [et al.] -- Habitats and benthos of a deep-sea marginal plateau, Lord Howe rise, Australia / Peter T. Harris ... [et al.] -- Seamounts, ridges, and reef habitats of American Samoa / Dawn J. Wright ... [et al.] -- Mapping condor seamount seafloor environment and associated biological assemblages (Azores, NE Atlantic) / Fernando Tempera ... [et al.] -- Cold-water coral colonization of Alboran sea knolls, western Mediterranean Sea / Ben De Mol ... [et al.] -- Fluid venting through the seabed in the Gulf of Cadiz (SE Atlantic Ocean, western Iberian Peninsula): geomorphic features, habitats, and associated fauna / José L. Rueda ... [et al.] -- Habitats or the Su Su Knolls hydrothermal site, eastern Manus Basin, Papua New Guinea / Yannick C. Beaudoin and Samantha Smith -- Southern Kermadec arc-Havre trough geohabitats and biological communities / Richard J. Wysoczanski and Malcolm R. Clark -- GeoHab atlas of seafloor geomorphic features and benthic habitats: synthesis and lessons learned / Peter T. Harris and Elaine K. Baker. Subjects: Ocean bottom--Maps. Benthos--Habitat--Maps. Environmental mapping--Case studies. Submarine topography--Maps--Case studies. Notes: Outgrowth of the 2007 GeoHab meeting held in Noumea, New Caledonia. Cf. pref. Includes bibliographical references. Series: Elsevier insights Elsevier insights.

Social networks and regional identity in Bronze Age Italy LCCN: 2014004594 Personal name: Blake, Emma. Main title: Social networks and regional identity in Bronze Age Italy / Emma Blake. Published/Produced: New York, NY: Cambridge University Press, 2014. Description: pages cm ISBN: 9781107063204 (hardback) 9781107636057 (paperback) LC classification: GN778.22.I8 B55 2014 Scope and content: "This book takes an innovative approach to detecting regional groupings in peninsular Italy during the Late Bronze Age, a notoriously murky period of Italian prehistory. Applying social network analysis to the distributions of imports and other distinctive objects, Emma Blake reveals previously unrecognized exchange networks that are in some cases the precursors of the named peoples of the first millennium BC: the Etruscans, the Veneti, and others. In a series of regional case studies, she uses quantitative methods to both reconstruct and analyze the character of these early networks and posits that, through path dependence, the initial structure of the networks played a role in the success or failure of the groups occupying those same regions in later times. This book thus bridges the divide between Italian prehistory and the Classical period, and demonstrates that Italy's regionalism began far earlier than previously thought"--Provided by publisher. Contents: Introduction: The problem of Italy's ancient peoples -- Imports and specialized products in Italy in the recent and final Bronze

Ages -- Group identity in prehistory: theory, interactions, and social networks -- The recent and final Bronze Age peninsular networks: assessing structure and cohesion -- The northern networks from the Terramare to the Veneto -- West central Italy: networks and neighbors -- Marche, Umbria, and the Apennine Mountain muddle -- Southern Italy: networks by land and by sea -- Conclusions and aftermath. Subjects: Bronze age--Italy. Prehistoric peoples--Italy. Social networks--Italy--History--To 1500. Regionalism--Italy--History--To 1500. Group identity--Italy--History--To 1500. Social archaeology--Italy. SOCIAL SCIENCE / Archaeology. Italy--Antiquities. Notes: Includes bibliographical references and index.

Social, economic and symbolic perspectives at the dawn of metal production LCCN: 2012451225 Main title: Social, economic and symbolic perspectives at the dawn of metal production / edited by Cecilia Conati Barbaro, Cristina Lemorini. Published/Created: Oxford: Archaeopress, 2012. Description: vi, 81 p.: ill., maps; 30 cm. ISBN: 9781407309620 (pbk.) 1407309625 (pbk.) LC classification: CC79.M47 S63 2012 Related names: Conati Barbaro, Cecilia. Lemorini, Cristina. Summary: "A collection of papers focusing on questions of Copper Age metallurgical contexts, outlining the importance of an integrated analysis of artefacts, considering pottery, metal, stone and osseous productions as inseparable aspects of economic and social choices." Contents: Introduction / Cecilia Conati Barbaro and Cristina Lemorini -- References -- First metals, last flints? The lithic production of central Italy at the end of the Neolithic. Some considerations / Cecilia Conati Barbaro -- The beginning of metallurgic production and the socioeconomic transformations of the Sardinian eneolithic / Maria Grazia Melis ... [et al.] -- Buried without metal: the role of lithic kit in Chalcolithic funerary contexts of the Marche Region (Central Italy) / Cristina Lemorini -- Producing for the dead, using while alive: lithic tools production and consumption in the late Neolithic of north-eastern Iberia / Xavier Terradas and Antoni Palomo -- A view from the mines. Flint exploitation in the Gargano (South-Eastern Italy) and socio-economic aspects of raw materials procurement at the dawn of metal production / Massimo Tarantini -- Functional variability and activity areas in prehistoric mining operations in Northern Chile / Diego S. Salazar ... [et al.]. Subjects: Metal-work, Prehistoric--Social aspects--Europe. Metal-work, Prehistoric--Social aspects--Chile. Metal-work, Prehistoric--Economic aspects--Europe. Metal-work, Prehistoric--Economic aspects--Chile. Copper age--Europe. Copper age--Chile. Excavations (Archaeology)--Europe. Excavations (Archaeology)--Chile. Europe--Antiquities. Chile--Antiquities. Notes: Includes bibliographical references. Series: BAR international series; 2372 BAR international series; 2372.

Stratigraphy and geology of volcanic areas LCCN: 2010000941 Main title: Stratigraphy and geology of volcanic areas / edited by Gianluca Groppelli, Lothar Viereck-Goette. Published/Created: Boulder, Colo.: Geological Society of America, 2010. Description: vii, 291 p.: ill. (some col.), maps (some col.); 28 cm.+ CD-ROM (4 3/4 in.) ISBN: 9780813724645 (pbk. + CD-ROM) 0813724643 (pbk. + CD-ROM) LC classification: QE522 .S787 2010 QE1 .G292 no. 464 Related titles: Supplementary materials to Stratigraphy and geology of volcanic areas. Related names: Groppelli, Gianluca. Viereck-Goette, Lothar. Contents: Stratigraphic approach to geological mapping of the late Quaternary volcanic island of Lipari (Aeolian archipelago, southern Italy) / Federico Lucchi, Claudio Antonio Tranne, and Piermaria Luigi Rossi -- A new geological map of Stromboli volcano (Tyrrhenian Sea, Italy) based on application of lithostratigraphic and unconformity-bounded stratigraphic (UBS) units / Alessandro Tibaldi -- Application of unconformity-bounded stratigraphic (UBS) units to the geological survey of the volcanic island Ustica (Italy) / Roberto Bonomo and Valeria Ricci -- Integrated approach for the reconstruction of stratigraphy and geology of Quaternary volcanic terrains: an application to the Vulsini Volcanoes (central Italy) / D.M. Palladino ... [et al.] -- Revised volcanostratigraphy of the Upper Miocene to Lower Pliocene Ürgüp Formation, Central Anatolian volcanic province, Turkey / L. Viereck-Goette ... [et al.] -- Resolving problems with the origin of Las Cañadas caldera (Tenerife, Canary Islands): Los Roques de García Formation--part of a major debris avalanche or an in situ, stratified, edifice-building succession? / Joan Martí ... [et al.] -- Ignimbrite stratigraphy and chronology on Terceira Island, Azores / Ralf Gertisser ... [et al.] -- The stratigraphic signature of volcanism off Campi Flegrei (Bay of Naples, Italy) / Alfonsa Milia -- Stratigraphy and volcanological evolution of the southwestern sector of Campi Flegrei and Procida Island, Italy / Annamaria Perrotta ... [et al.] -- Volcanological and structural evolution of the Ischia resurgent caldera (Italy) over the past 10 k.y. / Sandro de Vita ... [et al.] -- $^{40}Ar/^{39}Ar$ radiometric dating to constrain the volcanic stratigraphy: the Mt. Etna methodological case / Emanuela De Beni and Gianluca Groppelli -- Geologic mapping of the Colima volcanic complex (Mexico) and implications for hazard assessment / A. Cortés ... [et al.] -- Mapping and assessing volcanic and flood hazards and risks, with emphasis on lahars, in Arequipa, Peru / R.D. Vargas Franco ... [et al.] -- Probabilistic digital hazard maps for avalanches and massive pyroclastic flows using TITAN2D / Michael F. Sheridan ... [et al.]. Subjects: Volcanism. Volcanic fields. Geology, Stratigraphic. Form/Genre: Digital maps. Notes: Includes bibliographical references. Series: Special paper; 464 Special papers (Geological Society of America); 464.

Studies on freshwater Copepoda: a volume in honour of Bernard Dussart LCCN: 2011276041 Main title: Studies on freshwater Copepoda: a volume in honour of Bernard Dussart / by Danielle Defaye, Eduardo Suárez-Morales and J. Carel von Vaupel Klein (editors); with Victor Alekseev ... [et al.] (advisory editors). Published/Created: Leiden; Boston: Brill, c2011. Description: vii, 566 p.: ill. (some col.), maps; 25 cm. ISBN: 9789004181380 9004181385 LC classification: QL444.C7 S86 2011 QL435.A1 C753 no. 16 2011 Related names: Defaye, Danielle. Suárez, Eduardo, 1960- Vaupel Klein, J. C. von. Contents: Bernard Dussart, 17 August 1922-1 November 2008 / Danielle Defaye & J. Carel von Vaupel Klein -- Taxonomic differentiation and world geographical distribution of the Eucyclops serrulatus group (Copepoda, Cyclopidae, Eucyclopinae) / Victor R.

Alekseev & Danielle Defaye -- Brush setae on the antennules of male Eucyclopinae (Copepoda, Cyclopoida, Cyclopidae) / Geoff A. Boxshall & Terue C. Kihara -- Copepoda from a deep-groundwater porous aquifer in contact with karst: description of a new species, Paramorariopsis brigitae n. sp. (Copepoda, Harpacticoida) / Anton Brancelj -- Variazioni sul tema II: Description of Carcinocaris dussarti sp. nov., associated with xanthid crabs, and new data on the distribution of Carcinocaris serrichelata (Copepoda, Harpacticoida) / V. Cottarelli & M.C. Bruno -- Female genitalia in Cyclopidae (Copepoda, Cyclopoida): how to reinterpret the internal structures? / Corinne Cuoc & Danielle Defaye -- Copepodids of Pseudocyclops schminkei (Copepoda, Calanoida) and the relationship of Pseudocyclopidae to other calanoids / Frank D. Ferrari, Supawadee Chullasorn & Hans-Uwe Dahms -- Apalachocyclops gen. nov. (Copepoda, Cyclopoida, Cyclopidae): a Nearctic "Doppelgänger" of the European speocyclopines / Frank Fiers -- Patterns of copepod diversity in springs of central Italy (Copepoda: Cyclopoida, Harpacticoida): implications for conservation issues / Diana M.P. Galassi, Barbara Fiasca & Dina Del Tosto -- Morphology of Elaphoidella grandidieri (Guerne & Richard, 1893) (Copepoda, Harpacticoida) from Mexico with notes on fecundity in culture conditions / Martha A. Gutiérrez-Aguirre ... [et al.] -- Latitudinal gradients in diversity of the freshwater copepod family Cyclopidae / Maria Hołyńska -- Postnaupliar antennular development in the obligate subterranean Acanthocyclops kieferi (Chappuis, 1925) species-group (Kiefer, 1927) (Copepoda, Cyclopoida) / Sanda Iepure & Ioana Nicoleta Meleg -- Third genus of parastenocaridid copepods from Australia supported by molecular evidence (Copepoda, Harpacticoida) / T. Karanovic & S.J.B. Cooper -- Löffler's Chilean Eucyclops (Copepoda, Cyclopoida, Cyclopidae) revisited / Silvina Menu-Marque & Cecilia Locascio De Mitrovich -- Presumed parasitic trends in the evolution of the family Cyclopidae (Copepoda, Cyclopoida) / V.I. Monchenko -- Limnetic zooplankton of Chilean lakes and reservoirs: a tribute to Bernard Dussart / Vivian Montecino ... [et al.] -- Zooplankton communities of the

Sélingué Reservoir (Mali, West Africa). The role of environmental factors / M. Pagano, R. Arfi & D. Defaye -- A new species of the rare genus Idiodiaptomus Kiefer, 1936 (Copepoda, Calanoida, Diaptomidae) from northeastern Argentina / Juan C. Paggi -- Diversity and spatial distribution of copepods in the St. Lawrence River / B. Pinel-Alloul, E. Cusson & L. Aldamman -- Two new hyporheic Parastenocarididae from India: Parastenocaris sutlej n. sp. and P. gundlakamma n. sp. (Copepoda, Harpacticoida) / Yenumula Ranga Reddy -- Acanthocyclops magistridussarti sp. nov., from ground waters of peninsular Italy, with comments on the intraspecific variability of the antennary basis ornamentation (Copepoda, Cyclopoida, Cyclopidae) / Fabio Stoch & Maria Cristina Bruno -- Familial level phylogeny of free-living cyclopoids (Copepoda), inferred from partial 18S ribosomal DNA / Grace A. Wyngaard, Carlos E.F. Rocha & Almir Pepato -- Freshwater calanoids (Copepoda, Calanoida) of Taiwan with some comments on the morphology of Neutrodiaptomus tumidus Kiefer, 1937 observed by SEM / Shuh-Sen Young & C.-T. Shih. Subjects: Dussart, Bernard. Copepoda. Freshwater biology. Limnology. Notes: Includes bibliographical references. Abstracts also in French. Series: Crustaceana monographs, 1570-7024; 16 Crustaceana monographs; 16.

The best pasta sauces: favorite regional Italian recipes LCCN: 2014017531 Personal name: Negrin, Micol. Main title: The best pasta sauces: favorite regional Italian recipes / Micol Negrin. Edition: First edition. Published/Produced: New York: Ballantine Books, [2014] Description: xvi, 238 pages: color illustrations; 21 cm Links: Cover image 9780345547149.jpg ISBN: 9780345547149 (hardback) 9780345547156 (ebook) LC classification: TX819.A1 N44 2014 Summary: "Food writer, cooking teacher, and Milan native Micol Negrin has written the first cookbook to explore the best, most authentic Italian pasta sauce recipes from a regional perspective. The culinary odyssey begins in northern Italy, where rich sauces are prepared with fresh cream or local cheese: Creamy Fontina Sauce with Crushed Walnuts and White Truffle Oil; Parmigiano Sauce with Fresh Nutmeg; Pine Nut and Marjoram Pesto. Central Italy is known for sauces made with cured meats, sheep's milk cheeses, and extra-virgin olive oil: Spicy Tomato, Onion, and Guanciale Sauce; Crushed Potato Sauce with Cracked Black Pepper and Olive Oil; Caramelized Fennel and Crumbled Sausage Sauce. In southern Italy, simple, frugal ingredients meld into satisfying and delicious flavors: Sweet Pepper and Lamb Ragù with Rosemary; Fresh Ricotta Sauce with Diced Prosciutto; Spicy Cannellini Bean Sauce with Pancetta and Arugula. The islands of Sicily and Sardinia take advantage of the bountiful seafood from the Mediterranean and game from the mountainous terrain: Pork Ragù with a Hint of Dark Chocolate and Cinnamon; Red Mullet Roe Sauce with Garlicky Bread Crumbs; Rich Lobster Sauce. Negrin also provides a primer on saucing the Italian way, the basics for handmade pastas, the key to cooking pasta al dente, vital ingredients for every Italian kitchen, perfect wine pairings, and the best food shopping sources around the United States. Complete with mouthwatering color photographs and detailed maps of the various regions, The Best Pasta Sauces is like traveling to Italy without ever leaving your kitchen"-- Provided by publisher. Subjects: Sauces. Cooking, Italian. COOKING / Regional & Ethnic / Italian. COOKING / Specific Ingredients / Pasta. COOKING / Reference. Form/Genre: Cookbooks. Notes: Includes index.

The Cambridge prehistory of the Bronze and Iron Age Mediterranean LCCN: 2013047251 Main title: The Cambridge prehistory of the Bronze and Iron Age Mediterranean / [edited by] A. Bernard Knapp, Peter van Dommelen. Published/Produced: New York, NY: Cambridge University Press, 2015. Description: xviii, 688 pages: illustrations; 29 cm ISBN: 9780521766883 (hardback) LC classification: GN778.25 .C36 2015 Variant title: Prehistory of the Bronze and Iron Age Mediterranean Related names: Knapp, Arthur Bernard. Dommelen, Peter Alexander René van, 1966- Scope and content: "The Cambridge Prehistory of the Bronze and Iron Age Mediterranean offers new insights into the material and social practices of many different Mediterranean peoples during the Bronze and Iron Ages, presenting in particular those features that both connect and distinguish them. Contributors discuss in depth a range of topics that motivate and structure Mediterranean archaeology today, including insularity and connectivity; mobility, migration, and colonization; hybridization and cultural encounters; materiality, memory, and identity; community and household; life and death; and ritual and ideology. The volume's broad coverage of different approaches and contemporary archaeological practices will help practitioners of Mediterranean archaeology to move the subject forward in new and dynamic ways. Together, the essays in this volume shed new light on the people, ideas, and materials that make up the world of Mediterranean archaeology today, beyond the borders that separate Europe, Africa, and the Middle East"-- Provided by publisher. Contents: Mediterranean Introductions -- [Part I] Insularity and Connectivity -- 1. A little history of Mediterranean island prehistory / J.F. Cherry and T. Leppard -- 2. Inside out? Materiality and connectivity in the Aegean archipelago / C. Knappett and I. Nikolakopoulou -- 3. Early island exploitations: productive and subsistence strategies on the prehistoric Balearic Islands / D. Ramis -- 4. Islands and mobility: exploring Bronze Age connectivity in the South-Central Mediterranean / D. Tanasi and N. Vella -- 5. Sicily in Mediterranean history in the second millennium BC / A.M. Bietti Sestieri -- 6. Late Bronze Age Sardinia: Acephalous cohesion / E. Blake -- [Part II] Mobility, Migration and Colonization -- 7. Corridors and colonies: comparing fourth-third millennia BC interactions in Southeast Anatolia and the Levant / R. Greenberg and G. Palumbi -- 8. The Anatolian context of Philia material culture in Cyprus / C. Bachhuber -- 9. Bronze Age European elites: from the Aegean to the Adriatic, and back again / M. Galaty, H. Tomas and W.A. Parkinson -- 10. Greece in the Early Iron Age: mobility, commodities, polities, and literacy / J. Papadopoulos -- 11. Before "the gates of Tartessos": indigenous knowledge and exchange networks in the Late Bronze Age far west / M. Ruiz-Gálvez -- 12. Colonizations and cultural developments in the central Mediterranean / T. Hodos -- 13. The Iron Age in South Italy:

settlement, mobility, and culture contact / M. Osanna -- [Part III] Hybridisation and Cultural Encounters -- 14. Migration, hybridization, and resistance: identity dynamics in the Early Iron Age Southern Levant / S. Bunimovitz and Zvi Lederman -- 15. Cultural interactions in Iron Age Sardinia / C. Tronchetti -- 16. Myth into art: foreign impulses and local responses in archaic Cypriot sanctuaries / D. Counts -- 17. Mobility, interaction, and power in the Iron Age Western Mediterranean / J. Vives-Ferrándiz Sánchez -- [Part IV] Materiality, Memory and Identity -- 18. Sensuous memory, materiality, and history: rethinking the "rise of the palaces" on Bronze Age Crete / Y. Hamilakis -- 19. Beyond iconography: meaning-making in Late Bronze Age Eastern Mediterranean visual and material culture / M. Feldman -- 20. Changes in perceptions of the "other" and expressions of Egyptian self-identity in the Late Bronze Age / R.G. Roberts -- 21. The lure of the artifact? The effects of acquiring Eastern Mediterranean material culture / M. Kersel -- 22. Stone worlds: technologies of rock-carving and place-making in Anatolian landscapes / O. Harmanşah -- [Part V] Community and Household -- 23. Rethinking the late Cypriot built environment: households and communities as places of social transformation / K. Fisher -- 24. Households, hierarchies, territories, and landscapes in Bronze Age and Iron Age Greece / L. Foxhall -- 25. Connectivity beyond the urban community in central Italy / C. Riva -- 26. Long term social change in Iron Age northern Iberia (ca. 700-200 BC) / J. Sanmartí Grego -- 27. Who lives there? Settlements, houses, and households in Iberia / H. Bonet-Rosado and C. Mata-Parreño -- 28. Landscapes and seascapes of Southwest Iberia in the first millennium BC / A. Rodríguez Díaz -- 29. Domestic and settlement organization in Iron Age Southern France / M.C. Belarte -- [Part VI] Life and Death -- 30. Beyond the general and the particular: rethinking death, memory, and belonging in Early Bronze Age Crete / D. Catapoti -- 31. From the nineteenth century to the twenty-first: understanding the Bronze argaric lifecourse in the Mediterranean "far West" / S. Montón-Subías -- 32. Crossing borders: death and life in second millennium BC southern Iberia and North Africa / K. Lillios -- 33. An entangled past: island interactions, mortuary practices, and the negotiation of identities on Early Iron Age Cyprus / S. Janes -- 34. The violence of symbols: ideologies, identity, and cultural interaction in central Italian cemeteries / M. Cuozzo -- [Part VII] Ritual and Ideology -- 35. The Early Bronze Age Southern Levant: the ideology of an aniconic reformation / Y. Yekutieli -- 36. Ritual as the setting for contentious interaction: from social negotiation to institutionalized authority in Bronze Age Cyprus / J.M. Webb -- 37. Cult activities among central and north Italian protohistoric communities / A. Guidi -- 38. Ritual and ideology in Early Iron Age Crete: the role of the past and the East / M. Prent. Subjects: Bronze age--Mediterranean Region. Iron age--Mediterranean Region. Prehistoric peoples--Mediterranean Region. Material culture--Mediterranean Region--History--To 1500. Social archaeology--Mediterranean Region. Archaeology--Mediterranean Region. SOCIAL SCIENCE / Archaeology. Mediterranean Region--Antiquities. Notes: Includes bibliographical references and index.

The Carolingian world LCCN: 2010054602 Personal name: Costambeys, Marios. Main title: The Carolingian world / Marios Costambeys, Matthew Innes, Simon MacLean. Published/Created: Cambridge; New York: Cambridge University Press, 2011. Description: xix, 505 p.: ill., maps; 23 cm. Links: Cover image http://assets.cambridge.org/97805215/63666/cover/9780521563666.jpg ISBN: 9780521563666 (hardback) 0521563666 (hardback) 9780521564946 (pbk.) 0521564948 (pbk.) LC classification: DC70 .C67 2011 Related names: Innes, Matthew. MacLean, Simon. Summary: “At its height, the Carolingian empire spanned a million square kilometres of Western Europe--from the English Channel to central Italy and northern Spain, and from the Atlantic to the fringes of modern Hungary, Poland, and the Czech Republic. As the largest political unit for centuries, the empire dominated the region and left an enduring legacy for European culture. This long-awaited and comprehensive survey traces this great empire's history, from its origins around 700, with the rise to dominance of the Carolingian dynasty, through its expansion by ruthless military conquest and political manoeuvring in the eighth century, to the struggle to hold the empire together in the ninth. It places the complex political narrative in context, giving equal consideration to vital themes such as beliefs, peasant society, aristocratic culture, and the economy. Accessibly written yet authoritative, this book offers distinctive perspectives on a formative period in European history"-- Provided by publisher. Contents: The creation of Carolingian kingship to 800 -- Belief and culture -- Inventing the Carolingian empire: politics and government, 800-840 -- Villages and villagers, land and landowners -- Elite society -- Exchange and trade: the Carolingian economy -- Sustaining the Carolingian empire: politics and government, 840-888 -- Epilogue. Subjects: Carolingians--History. Civilization, Medieval. France--History--To 987. France--Civilization. Europe, Western--History. Europe, Western--Civilization. Europe--History--476-1492. Notes: Includes bibliographical references and index. Series: Cambridge medieval textbooks Cambridge medieval textbooks.

The Colli Albani volcano LCCN: 2012360574 Main title: The Colli Albani volcano / edited by R. Funiciello and G. Giordano. Published/Created: London: The Geological Society, 2010. Description: vi, 392 p: ill., maps; 25 cm + 1 folded col. map. ISBN: 9781862393073 1862393079 LC classification: QE526.2.I8 C65 2010 Related names: Funiciello, R. Giordano, G. (Guido) International Association of Volcanology and Chemistry of the Earth's Interior. Contents: The Colli Albani volcano: foreword and previous studies / R. Funiciello & G. Giordano -- The Tyrrhenian margin geological setting: from the Apennine orogeny to the K-rich volcanism -- M. Mattei, S. Conticelli, & G. Giordano -- The three-dimensional structure of the Colli Albani volcano / C. Chiarabba ... [et al.] -- Stratigraphy, volcano tectonics and evolution of the Colli Albani volcanic field / G. Giordano & the CARG Team -- Geochronology of Colli Albani volcano / M. Soligo & P. Tuccimei -- Geochemistry, isotopes and mineral chemistry of the Colli Albani volcanic rocks: constraints on magma genesis and evolution / S. Conticelli ... [et al.] -- The sedimentary substrate of the

Colli Albani volcano / E. Danese & M. Mattei -- Metamorphic, metasomatic and intrusive xenoliths of the Colli Albani volcano and their significance for the reconstruction of the volcano plumbing system / A. A. De Benedetti ... [et al.] -- Three-dimensional reconstruction of the main unconformities of the Colli Albani stratigraphy and deposit volume calculations / G. Diano, A. Bonamico, & G. Giordano -- Hydrogeology of the Colli Albani volcano / R. Mazza & G. Capelli -- Origin and deposition of the Lapis Tiburtinus travertine / C. Faccenna, R. Funiciello, & M. Soligo -- Lake Albano: bathymetry and level changes / M. Anzidei & A. Esposito -- Reconstructing natural and human-induced environmental change in central Italy since the late Pleistocene: the multi-proxy records from maar lakes Albano and Nemi / L. Vigliotti ... [et al.] -- Geochemistry of the Albano crater lake / M. L. Carapezza, M. Lelli, & L. Tarchini -- Physico-chemical and biological aspects of Lake Albano / N. T. W. Ellwood ... [et al.] -- Volcanic hazards of the Colli Albani / M. L. Carapezza ... [et al.] -- Current geodetic deformation of the Colli Albani volcano: a review / M. Anzidei, F. Riguzzi, & S. Stramondo -- Modelling of CO_2 circulation in the Colli Albani area / M. Todesco & G. Giordano -- Volcanic activity of the Lake Albano maar in Roman history and mythology / R. Funiciello ... [et al.] -- Geomorphological and environmental transformations during the recent prehistory: a reconstruction of the landscape and the peopling of the territory southeast of Rome / A. P. Anzidei ... [et al.] -- Archaeological investigations in the Torre Spaccata valley (Rome): human interaction with the recent activity of the Albano maar / P. Gioia ... [et al.]. Subjects: Volcanoes--Italy--Alban Hills. Volcanoes--Italy--Rome (Province) Geology--Italy--Alban Hills. Notes: Includes bibliographical references and index. Series: Special publications of the International Association of Volcanology and Chemistry of the Earth's Interior; no. 3 Special publications of the International Association of Volcanology and Chemistry of the Earth's Interior; no. 3.

The genealogy of 4 little known families from central Italy: the Fazzini and the Pignotti, the Bernardini and the Galdenzi (1540-2015) LCCN:

2015946626 Personal name: Fazzini, Paul Francis. Main title: The genealogy of 4 little known families from central Italy: the Fazzini and the Pignotti, the Bernardini and the Galdenzi (1540-2015) / Paul Francis Fazzini. Published/Produced: Chicago, IL: Society of Paul Francis Fazzini, 2015. Description: pages cm ISBN: 9780692489802

The Hellenistic West: rethinking the ancient Mediterranean LCCN: 2013013369 Main title: The Hellenistic West: rethinking the ancient Mediterranean / edited by Jonathan R.W. Prag and Josephine Crawley Quinn. Published/Produced: Cambridge; New York: Cambridge University Press, 2013. Description: xxi, 471 pages: illustrations, maps; 25 cm Links: Contributor biographical information http://www.loc.gov/catdir/enhancements/fy1310/2013013369-b.html Publisher description http://www.loc.gov/catdir/enhancements/fy1310/2013013369-d.html Table of contents only http://www.loc.gov/catdir/enhancements/fy1310/2013013369-t.html ISBN: 9781107032422 LC classification: DF235 .H45 2013 Related names: Prag, J. R. W., author, editor of compilation. Quinn, Josephine Crawley, author, editor of compilation. Contents: Introduction / Jonathan R.W. Prag and Josephine Crawley Quinn -- The view from the East / Andrew Erskine -- Hellenistic Pompeii: between Oscan, Greek, Roman, and Punic / Andrew Wallace-Hadrill -- The 'Hellenistics of death' in Adriatic central Italy / Edward Bispham -- Hellenistic Sicily, c.270-100 BC / R.J.A. Wilson -- Trading across the Syrtes: Euesperides and the Punic world / Andrew Wilson -- Strangers in the city: elite communication in the Hellenistic central Mediterranean / Elizabeth Fentress -- Monumental power: 'Numidian royal architecture' in context / Josephine Crawley Quinn -- Representing hellenistic Numidia, in Africa and at Rome / Ann Kuttner -- Hellenism as subaltern practice: rural cults in the Punic world / Peter van Dommelen and Mireia Lopez-Bertran -- Were the Iberians Hellenized? / Simon Keay -- Epigraphy in the western Mediterranean: a Hellenistic phenomenon? / Jonathan R.W. Prag -- Heracles, coinage, and the West: three Hellenistic case-studies / Liv Mariah Yarrow -- On the significance of East and West in today's 'Hellenistic' history / Nicholas Purcell. Subjects: Hellenism--History. Greeks--Mediterranean Region--History. Greeks--Colonization--Mediterranean Region. Mediterranean Region--Civilization--Greek influences. Mediterranean Region--History--To 476. Notes: Includes bibliographical references (pages 391-459) and index.

The mysterious and unknown rulers of central Italy in the Early Middle Ages LCCN: 2011941342 Personal name: DeMay, Jeffrey M. Main title: The mysterious and unknown rulers of central Italy in the Early Middle Ages / Jeffrey M. DeMay. Published/Created: Pace, FL: Trents Prints, 2011. Description: p. cm. ISBN: 9781937000127

The Roman crucible: the artistic patronage of the Papacy, 1198-1304 LCCN: 2013482454 Personal name: Gardner, Julian, author. Main title: The Roman crucible: the artistic patronage of the Papacy, 1198-1304 / Julian Gardner. Published/Produced: München: Hirmer Verlag, [2013] ©2013 Description: 516 pages: illustrations; 31 cm. Links: Publisher description http://www.loc.gov/catdir/enhancements/fy1506/2013482454-d.html Table of contents only http://www.loc.gov/catdir/enhancements/fy1506/2013482454-t.html Contributor biographical information http://www.loc.gov/catdir/enhancements/fy1506/2013482454-b.html ISBN: 9783777423852 3777423858 LC classification: N5273 .G372 2013 Portion of title: Artistic patronage of the Papacy, 1198-1304 Contents: The preconditions of patronage -- The curia in the thirteenth century -- The earthly frame: architecture in Rome -- Novus ordo: the new architectural style in Rome and central Italy -- French patrons and Italian art in the thirteenth century -- The revival of monumental sculpture in Rome -- Submerged evidence: texts and wreckage -- Roman readers -- English embroiderers -- Mosaic and painting in Rome 1198-1276: Innocent III to Gregory X -- Pope Nicholas III and the birth of modern painting in Rome -- The final flowering: Roman painting before Avignon -- Other Romes: the hill-stations -- Romans and Greeks: the relations of the Papacy and Byzantium during the thirteenth century -- Onera romanorum -- Home thoughts from abroad. Subjects: Art patronage--Italy--History--13th century. Popes--

Art patronage--History--13th century. Notes: Includes bibliographical references (pages 423-496) and index. Series: Römische Forschungen der Bibliotheca Hertziana; Band XXXIII Römische Forschungen der Bibliotheca Hertziana; Bd. 33.

TRAC 2012: proceedings of the twenty-second annual Theoretical Roman Archaeology Conference which took place at Goethe University in Frankfurt 29 March-1 April 2012 LCCN: 2013007802 Meeting name: Theoretical Roman Archaeology Conference (22nd: 2012: Universität Frankfurt am Main) Main title: TRAC 2012: proceedings of the twenty-second annual Theoretical Roman Archaeology Conference which took place at Goethe University in Frankfurt 29 March-1 April 2012 / edited by Annabel Bokern, Marion Bolder-Boos, Stefan Krmnicek, Dominik Maschek and Sven Page. Published/Produced: Oxford, UK: Oxbow Books; Oakville, CT: David Brown Book Company, 2013. Description: xi, 200 pages: illustrations; 25 cm ISBN: 9781782971979 LC classification: DG12.5 .T44 2012 Variant title: Theoretical Roman Archaeology Conference 2012 Related names: Bokern, Annabel, editor of compilation. Contents: Marks of imitation or signs of originality?: an approach to structural -- Supports in Roman marble statuary / Anna Anguissola -- Equites and senators as agents of change: urban culture and elite self-representation in Thamugadi and Lepcis Magna (second-third centuries A.D.) / Lennart Gilhaus -- Sacra volsiniensia: civic religion in Volsinii after the Roman conquest / Annalisa Calapà -- The internal frontier: an African model for culture change in south central Italy (fourth-third centuries B.C.) / Roman Roth -- Street activity, dwellings and wall inscriptions in ancient Pompeii: a holistic study of neighbourhood relations / Eeva-Maria Viitanen, Laura Nissinen and Kalle Korhonen -- Understanding neighbourhood relations through shared structures: reappraising the value of insula-based studies / Heini Ynnilä -- Secondary doors in entranceways at Pompeii: reconsidering access and the "view from the street" / Evan Proudfoot -- The perception of Egypt in networks of being and becoming: a thing theory approach to Egyptianising objects in Roman domestic contexts / Eva M. Mol -- Hybrid bridges: an exploration into how traditionally "Romanised" elements of the town interacted with meaning-laden pre-historic waterscapes / Jay Ingate -- Romanisation in Gaul: new methodological approaches for the study of Gaulish fine wares (200 B.C.-A.D. 50) / Sylvie Barrier -- Transgendered archaeology: the Galli and the Catterick transvestite / Renato Pinto and Luciano C.G. Pinto -- Ritual landscapes of pre-Roman Britain: the margins of practice on the margins of the empire / Nicky Garland. Subjects: Romans--Europe--Congresses. Archaeology--Philosophy--Congresses. Europe--Antiquities, Roman--Congresses. Rome--Antiquities--Congresses. Notes: Includes bibliographical references.

Tuscany and Umbria: the collected traveler LCCN: 2010010183 Main title: Tuscany and Umbria: the collected traveler / edited by Barrie Kerper. Published/Created: New York: Vintage Books, 2010. Description: xxiii, 616 p.: ill., map; 21 cm. Links: Contributor biographical information http://www.loc.gov/catdir/enhancements/fy1110/2010010183-b.html Publisher description http://www.loc.gov/catdir/enhancements/fy1110/2010010183-d.html Sample text http://www.loc.gov/catdir/enhancements/fy1110/2010010183-s.html Table of contents only http://www.loc.gov/catdir/enhancements/fy1110/2010010183-t.html ISBN: 9780307474902 0307474909 LC classification: DG734.23 .T87 2010 Related titles: Central Italy, Tuscany & Umbria. Related names: Kerper, Barrie. Contents: Italy -- My Italy / by Erica Jong -- Welcome to our rented nightmare / by Mary-Lou Weisman -- Buon viaggio: a bouquet of reminders / by Kate Simon -- Tuscany -- The trouble with Tuscany / by Fred Plotkin -- Market day in a Tuscan town / by Frances Mayes -- Italy's best-kept secret / by David Leavitt -- Only in Tuscany / by Dan Hofstadter -- Coming home to chianti / by Gini Alhadeff -- Lucca: a Tuscan treasure / by Lorraine Alexander -- Siena in three acts / by William Zinsser -- Interview: Barbara Milo Ohrbach -- Interview: Frances Mayes -- Interview: Charlie Conrad -- Umbria -- Perugia / by Nadia Stancioff -- Spoleto: a town for all seasons / by Nadia Stancioff -- The hills of the sublime / by G.Y. Dryansky -- Interview:

Joan and Roger Arndt -- Florence -- The first time I saw Florence / by Sallie Tisdale -- The other side of the Arno / by Jo Durden-Smith -- Florence, then and now / by Adam Begley -- Learning to live with arrivederci / by Susan Jacoby -- Interview: Lisa McGarry -- La cucina Italiana -- Beaneaters and bread soup / by Lori De Mori -- Italy's original garlic bread / by S. Irene Virbila -- Tuscan olive oils / by Faith Heller Willinger -- Italy's vin santo: a sip of hospitality / by S. Irene Virbila -- Nach Waxman and Matt Sartwell favorite food books: an interview -- Interview: Sergio Esposito -- A tavola! -- Florentine trattorias / by Faith Willinger -- Florence: a restaurant renaissance / by Faith Willinger -- Valle del Serchio and the Garfagnana / by Beth Elon -- Interview: Faith Willinger -- Piazze, gardini, e monumenti -- Back roads of Tuscany / by William Sertl -- Botticelli's Primavera / by Roy McMullen -- The cloisters of Florence / by Louis Inturrisi -- The Last Supper, seen six ways / by Louis Inturrisi -- A Tuscan and Umbrian miscellany. Subjects: Travelers' writings. Interviews--Italy--Tuscany. Interviews--Italy--Umbria. Cooking, Italian. Tuscany (Italy)--Description and travel--Sources. Umbria (Italy)--Description and travel--Sources. Tuscany (Italy)--Biography. Umbria (Italy)--Biography. Tuscany (Italy)--Social life and customs--Sources. Umbria (Italy)--Social life and customs--Sources. Notes: "An inspired companion guide." "A Vintage departures original"--T.p. verso. Previous ed.: New York: Three Rivers Press, 2000, with title Central Italy, Tuscany & Umbria. Includes bibliographical references. Series: Vintage departures Vintage departures.

War, massacre, and recovery in Central Italy, 1943-1948 LCCN: 2010513632 Personal name: Belco, Victoria, 1952- Main title: War, massacre, and recovery in Central Italy, 1943-1948 / Victoria C. Belco. Published/Created: Toronto: University of Toronto Press, c2010. Description: xi, 574 p., [16] p. of plates: ill., maps, ports.; 24 cm. ISBN: 9780802093141 (bound) 0802093140 (bound) LC classification: DG572 .B37 2010 Summary: "The Second World War wreaked unprecedented devastation throughout Europe, necessitating monumental reconstruction efforts that burdened not only governments, but the lives of ordinary citizens. War, Massacre, and Recovery in Central Italy, 1943-1948 examines this transitional period in the province of Arezzo by detailing the daily experiences of civilians through the traumas of war and the difficulties of recovery. Studying the aftermath of war in a new and insightful way, Victoria C. Belco shifts the perspective from the national to the local level. With this localized focus, she provides valuable insight into the ways in which civilians coped with an overwhelming range of problems - from adjusting to Allied occupation and widespread displacement to rampant unemployment and the restructuring of local administrations and institutions after fascism. Recreating the post-war atmosphere of disorder, need, and political upheaval, Belco shows how the competing community interests caused social fragmentations that impeded change, while the unity of a shared past prevented civil war."--pub. desc. Contents: Introduction: from war to peace -- Situating the villages -- Italy at war -- Variations on a massacre -- Completing the disintegration -- Life with the Allies -- True victims and the truly needy -- Restitution, reparations, and rewards -- Restoring the community -- Rebuilding the commune -- Private property, public good, and the housing crisis -- Mezzadria struggle and social change: the 'haves' and the 'have-nots' -- Unemployment -- Pure politics -- Honouring the dead -- Conclusion: after war and massacre. Subjects: World War, 1939-1945--Atrocities--Italy, Central. Civilians in war--Italy, Central--History--20th century. Reconstruction (1939-1951)--Italy--Citizen participation. Guerre mondiale, 1939-1945--Atrocités--Italie (Centre) Guerre--Participation des civils--Italie (Centre)--Histoire--20e siècle. Reconstruction, 1939-1951--Italie--Participation des citoyens. Italy--History--German occupation, 1943-1945. Italy--History--1945-1976. Italie--Histoire--1943-1945 (Occupation allemande) Italie--Histoire--1945-1976. Notes: Includes bibliographical references (p. [513]-537) and index. Series: Toronto Italian studies

INDEX

#

A

B

C

D

E

F

G

Q

R

S

T

U

V

W

Y

Z